Learn to Use Microsoft Access 2016

Michelle N. Halsey, PMP, CSM

ISBN-10: 1-64004-260-1

ISBN-13: 978-1-64004-260-5

TABLE OF CONTENTS

5

6

Chapter 1: Opening and Closing Access

Welcome to Learn How to Use Access 2016. In this chapter, you will learn how to open Access. You will learn about opening files, including from the Recent list. You will also gain an understanding of security warnings when working with databases. We will look at the Access desktop interface, including looking at the Ribbon and the Status bar. You will also learn about your Microsoft account. Finally, we will look at the difference between closing files and closing Access.

This guide will use the Northwind Sample Database. You can download this database using the following procedure:

Step 1: Click File and select New.

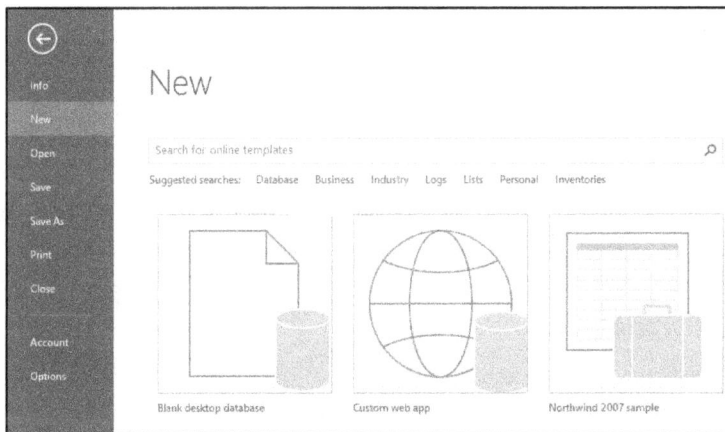

Step 2: Search for "Northwind" (without quotes).

Step 3: Click the Northwind 2007 Sample database.

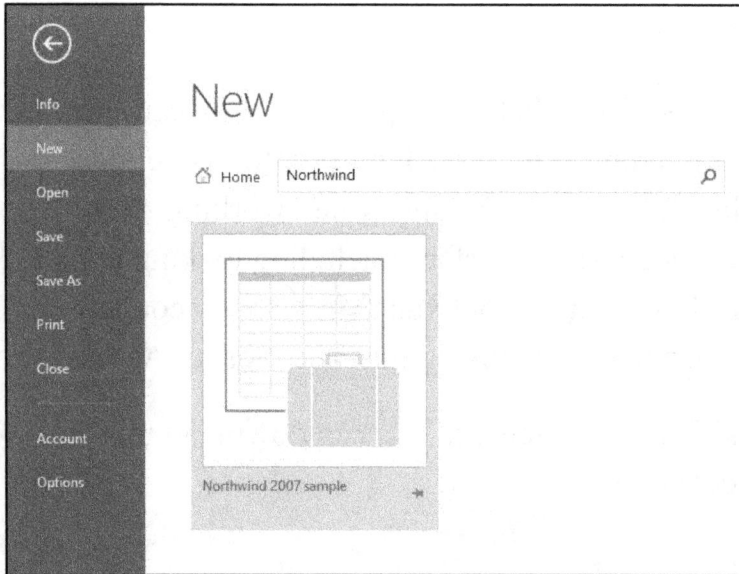

Step 4: Give the database a name and click Create.

You may see a message box that states "Please wait while Microsoft Access prepares the template for use."

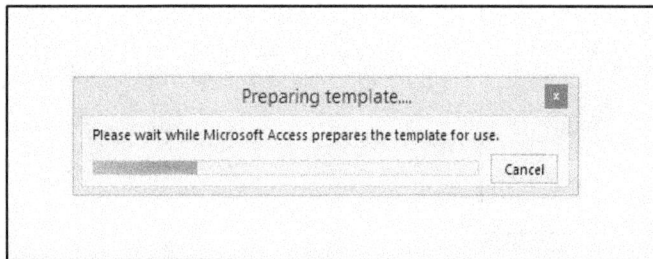

Step 5: Click Enable Content when the file opens to begin working with the database.

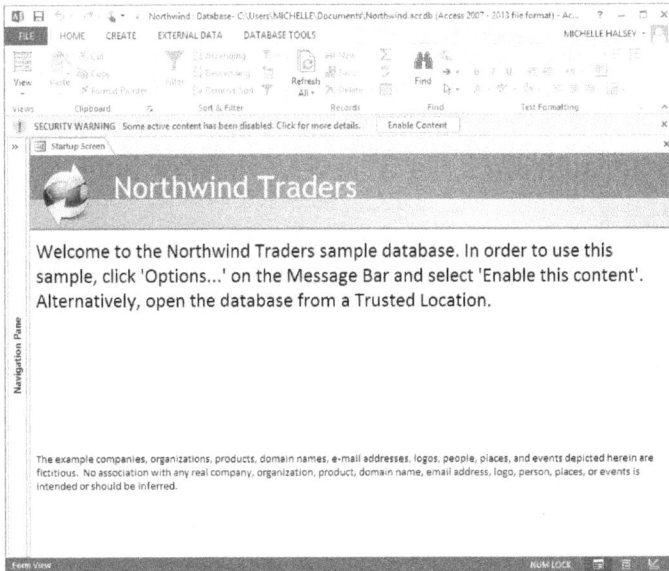

Opening Access

Let's open Access and get started.

To open Access in Windows 8 or Windows 10, use the following procedure.

Step 1: From the Start page, select the Access 2016 icon.

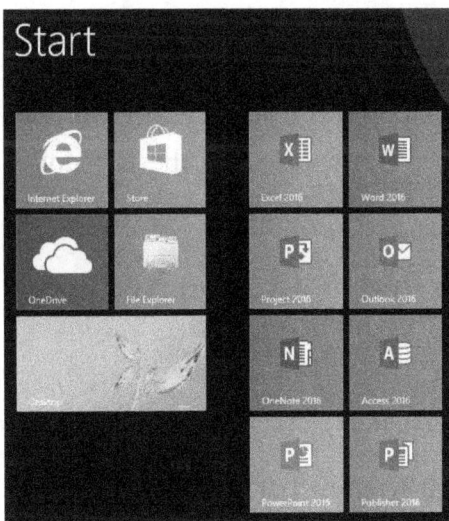

Use this procedure if using Windows 7:

Step 1: Select the Start icon from the lower left side of the screen.

Step 2: Select All Programs.

Step 3: Select Microsoft Office.

Step 4: Select Microsoft Office Access 2016.

Using the Recent List and Opening Files

When you first open Access, you can choose from a list of recent files, or open a different file.

To open a database from the Recent list, use the following procedure.

Select the database that you want to open from the Recent list.

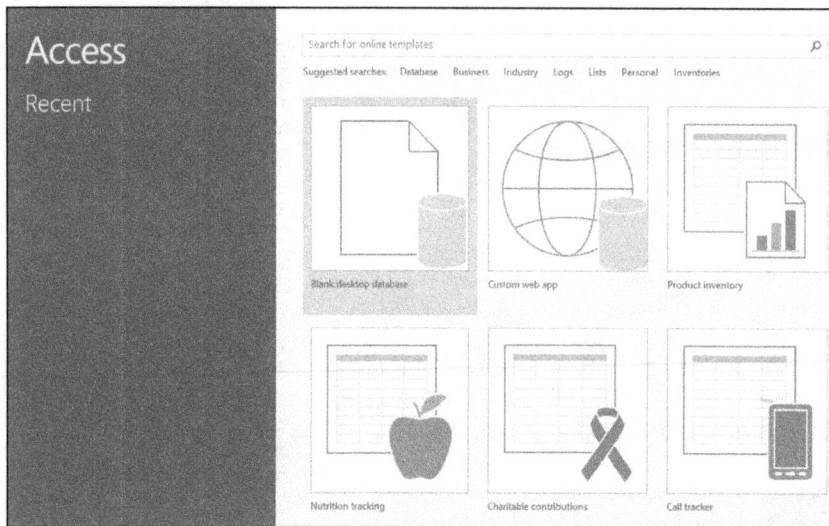

To pin an item on the Recent list, use the following procedure, click the pin on the right side of the Recent list item. The item moves to the top section of the Recent list.

To unpin an item, click the pin on the right side of the Recent list again. The item returns to the previous location in the Recent list.

To open a database, use the following procedure.

Step 1: Select Open Other Files from the bottom of the Recent list. Or select Open from the Backstage View.

Step 2: Select one of the Places you would like to look for the document. The default options are Recent, your Microsoft OneDrive location, Computer, or Add a Location.

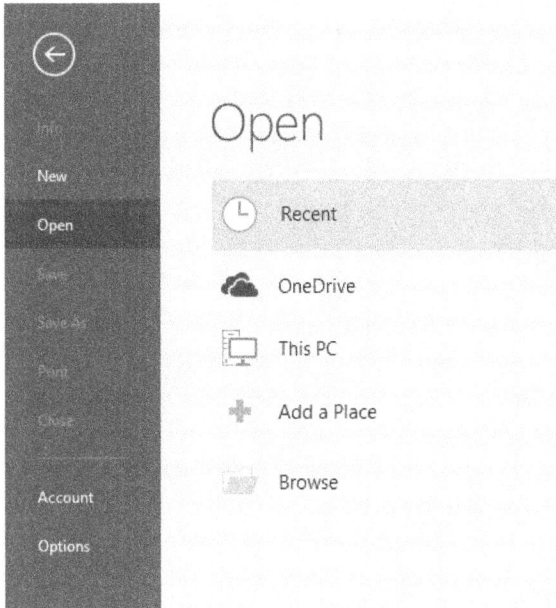

Open Database from OneDrive or on Computer

Step 1: To open a database from the OneDrive or your computer, select Browse.

Step 2: In the Open dialog box, navigate to the location of the file you want to open. Select it and select Open.

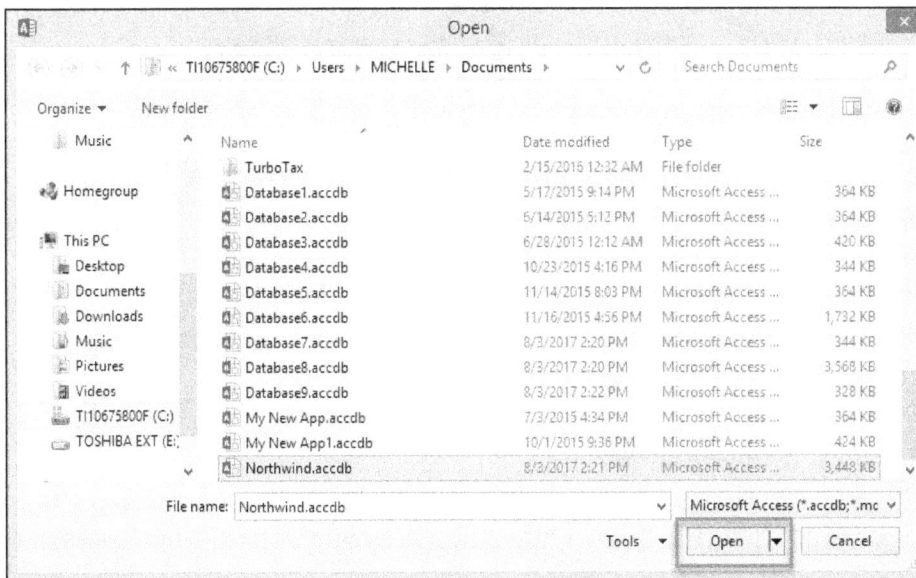

Understanding Security Warnings

The Message Bar displays security warnings when there is potentially unsafe active content in the file you want to open.

Review the Security Warning.

Understanding the Access Interface

Review the Access interface, including the Ribbon, the Navigation pane, the database window, the Quick Access toolbar, and the Status bar.

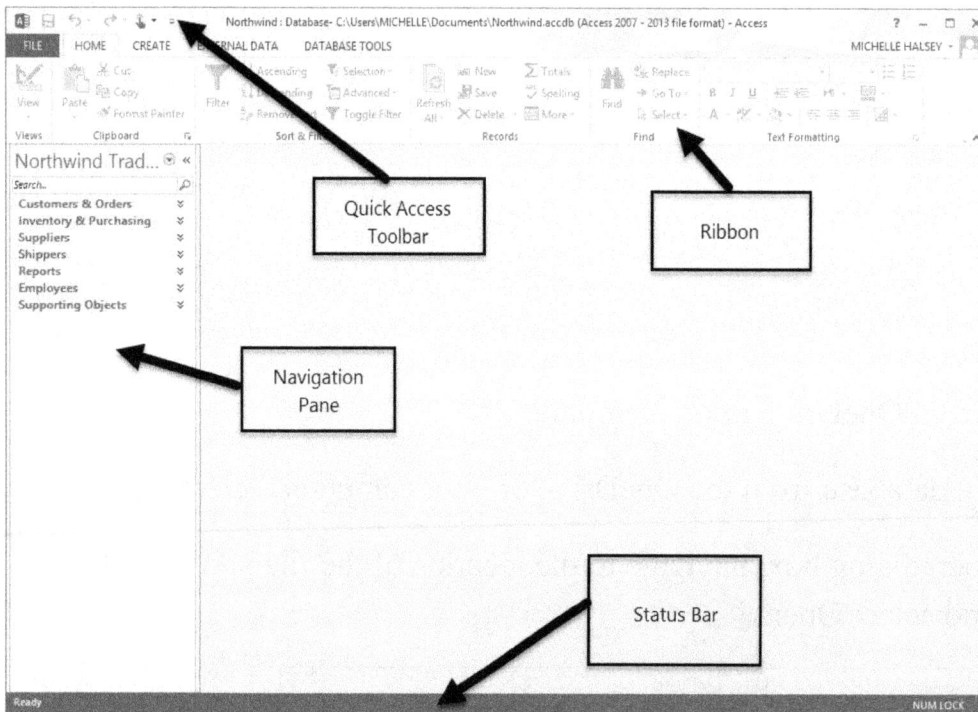

Each Tab in the Ribbon contains many tools for working with your database. To display a different set of commands, click the Tab name. Buttons are organized into groups according to their function.

The Quick Access toolbar appears at the top of the Access window. It provides you with one-click shortcuts to commonly used functions, like save, undo, and redo.

The Navigation pane allows you to work with your Database objects. We will take a closer look at the Navigation pane in Module Three.

The Status bar shows various information, depending on your activity in Access.

To collapse and un-collapse the Ribbon options, use the following procedure.

Step 1: Select the small arrow on the right side of the Ribbon.

While the Ribbon is collapsed, just click a Ribbon tab to activate it. When you select a command, the Ribbon is minimized again.

Step 2: To return to the Ribbon, right click one of the tabs.

Step 3: Select Collapse the Ribbon from the context menu.

About Your Account

Working in the cloud means that your files can be accessible from many different computers or devices. Microsoft manages that by linking Access to your Microsoft account.

To review the account options, perform the following:

Step 1: Click the file ribbon.

Step 2: Click Account.

Step 3: Sign in to make changes.

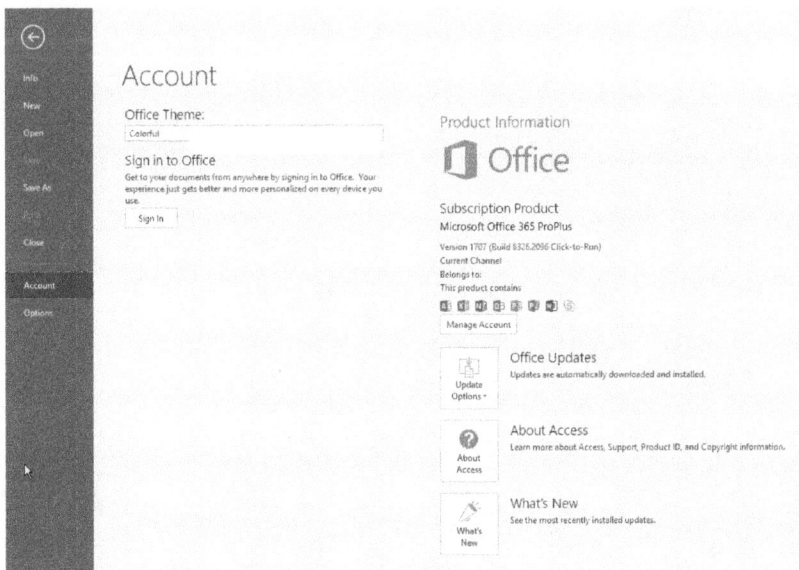

Closing Files vs. Closing Access

Our final lesson in the introduction to the Access interface is how to close database files and how to close the application.

To close a file, use the following procedure.

Step 1: Select the File tab from the Ribbon.

Step 2: Select Close from The Backstage View.

To close the application (if only one database is open), click the X at the top right corner of the window.

Chapter 2: An Introduction to Databases

Access is a relational database. A relational database is a collection of data items organized as a set of tables. In this chapter, you will get a chance to familiarize yourself with the basic components of the database. First, we will look at some common database terms. You will also take a closer look at the Navigation pane. We will introduce tables, table relationships, queries, forms, and reports. Finally, you will learn how to close database objects in Access.

About Common Database Terms

Before you start learning more about Access procedures, let's make sure that you have a solid grounding in database terminology.

The following terms are commonly used in Access:

- File – A file is a collection of associated records.
- Record – All information (all fields/columns) for every item in a file is called a record (or each individual line).
- Field – A record is divided into separate headings/sections and each is known as a field – this could refer to each column/heading. There are different types of fields, including:
    - NUMBER fields, which can be sorted in ascending or descending numeric order
    - Currency and Date/Time fields
    - TEXT fields, which can contain numbers and text that do not need to be sorted, such as telephone numbers
- Data – Data is a collection of pieces of information.
- Database – A database is the organized collection of your data. The Access database can be sorted, queried, or amended at any time.
- Database object – An object is a container for the work you want Access to perform. It includes tables, macros, queries, forms, reports, and/or pages.
- Table – A table is a collection of data organized by categories called fields, into unique sets of data called records.
- Datasheet – A datasheet is a different way of looking at a table, form query, or stored procedure. It is displayed in rows and columns.

- Query – A query is a request you make of your data to extract only the information you want.
- Form – A form is a user-friendly interface used for entering or displaying data.
- Report – A report is like a form, but it only shows the information you want. It is also the result of a query. You can print a report.

Using the Navigation Pane

The Navigation Pane provides a way to open all the database objects that make up your database.

To expand or collapse the Navigation pane, and how to expand or collapse each of the items listed in the pane.

To open the Navigation pane, select the bar on the left of the screen, titled "Navigation Pane."

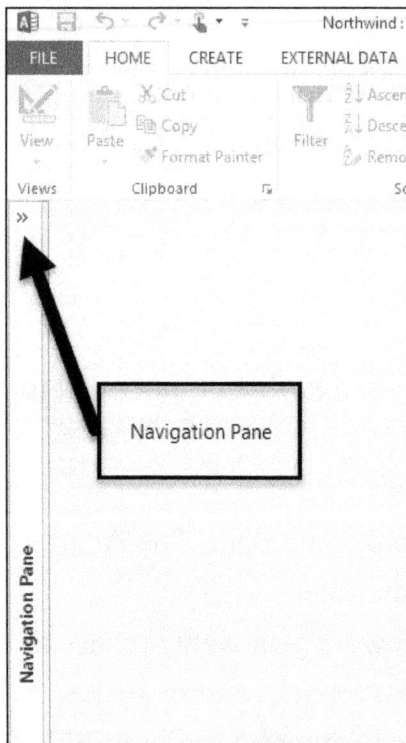

The Navigation Pane displays all the objects included with your database, including tables, forms, reports, and queries. The sample database is organized with custom categories.

Step 1: Select a Category header to expand it (or collapse it).

Step 2: Double-click an object to open it.

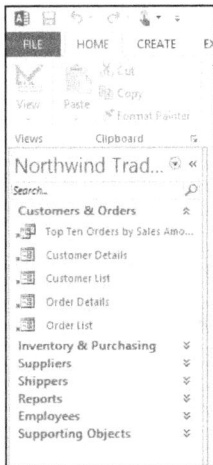

You can customize the Navigation Pane.

Step 3: Select the arrow in the Navigation Pane title bar.

Access displays a list.

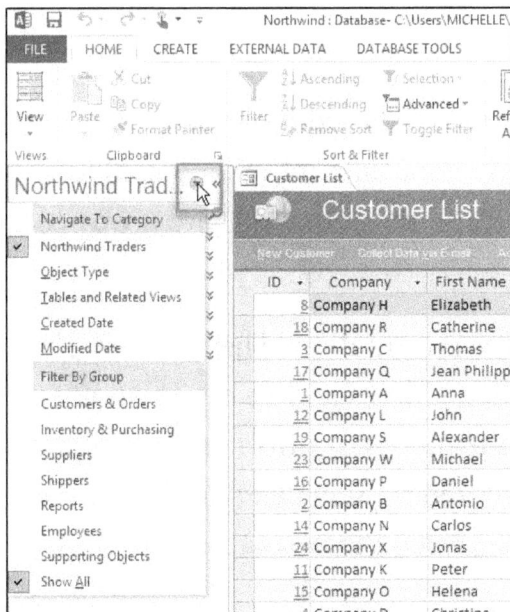

Step 4: Under the Navigate To Category, select the category that matches the objects you wish to view.

You can also filter the contents of the Navigation Pane. Select an option under Filter By Group to show only those items.

Understanding Tables and Table Relationships

Tables make up the backbone of your database. Tables have relationships to connect data without having to store it in multiple places.

Tables

Tables are made up of columns and rows. Each column is a field. A field is a single piece of information. For example, a field could be Last Name, or Product description. Each field includes some type of information relevant to the table you have created.

Each row is a record. A record is a meaningful and consistent way to combine information about something. In other words, the record includes all the fields and their information about one unique item in the table. The same fields of information are stored for each record in the table. Each record includes a unique ID known as the primary key.

Table Relationships

One of the goals of good database design is to remove data redundancy or duplicate data. To achieve this goal, you can divide your data into many subject-based tables so that each fact is only represented once. Relationships are a way to bring the divided information back together.

For example, in one table you could store customer information. The fields might include the customer number, first name, last name, address and so on. In another table, you might store order information. One of the pieces of information you may want to store about orders is the customer who ordered it. You can simply use the customer number, and then create a relationship between the two tables to access the rest of the customer information.

To open a table, use the following procedure.

Step 1: In the Navigation pane, select the arrow next to Northwind Traders.

Step 2: Select the Table Object Type from the drop-down list.

Step 3: Select the double arrows to the right of Tables. Double-click any table to open it.

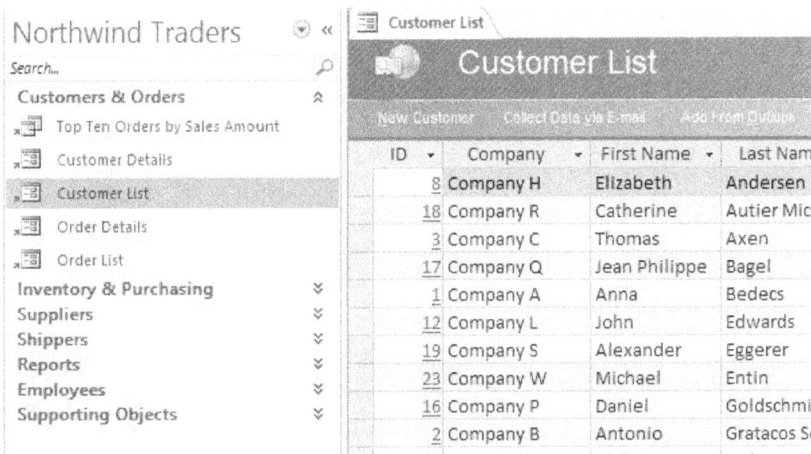

Notice the icon next to the tables.

Understanding Queries

Queries can ask a question or perform an action on your data. Queries and tables work together to make sense of your data.

To open a query, use the following procedure.

Step 1: Select the arrow in the Navigation Pane title bar. Access displays a list.

Step 2: Select the Queries Object type under Navigate To category.

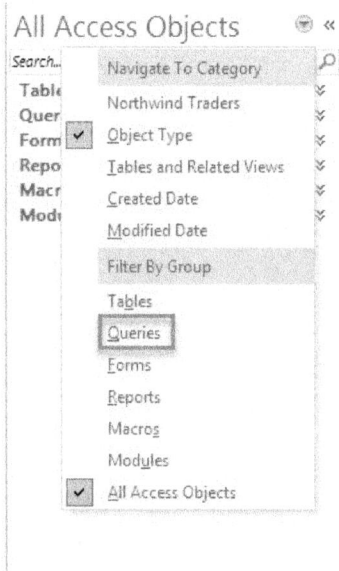

Step 3: In the Navigation pane showing types of objects, select the double arrows to the right of Queries. Double-click any query to open it.

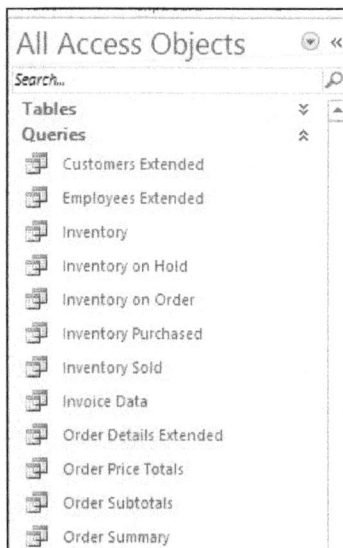

Notice the icons next to the queries.

Open a Report

Step 1: Select the arrow in the Navigation Pane title bar. Access displays a list.

Step 2: Select the Reports Object Type under the Navigate To category.

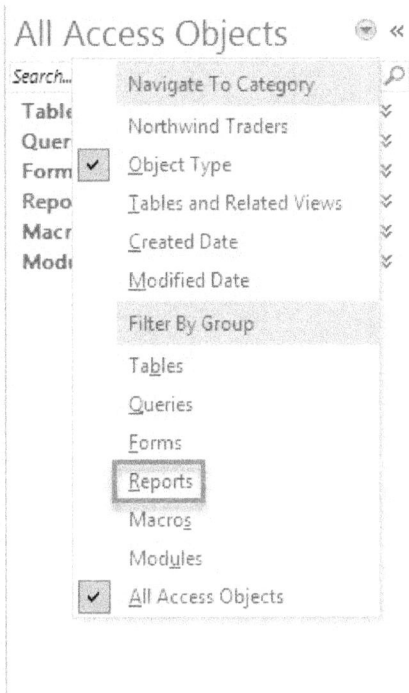

Step 3: Click the arrow next to Reports to expand the Reports category.

Step 4: Double-click the report to open the report.

Closing Database Objects

To close database objects, use the following procedure.

Step 1: Right-click the tab for the database object you want to close.

Step 2: Select Close from the context menu. You can also select Close All to close all open objects.

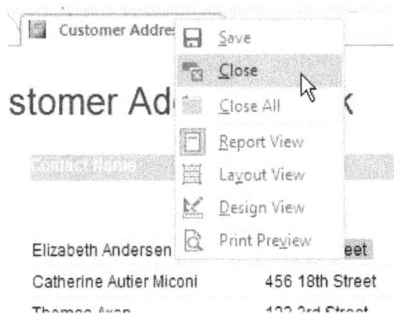

Chapter 3: An Introduction to Desktop Databases

In this chapter, we will switch back to the desktop database. You will learn how to open a table in Datasheet view. You will learn about Keys, Data Types, and Fields. We will also look at searching for records. Finally, you will learn how to delete a record.

Working with a Table in Datasheet View

The Datasheet view looks very much like a spreadsheet, with records in rows and fields in columns.

To open a table in datasheet view, when you double-click a table in the Navigation pane, it automatically opens in datasheet view.

Review the Table Tools tabs that appear when working with tables in Datasheet view. Hover over the commands to see ScreenTips.

Table Tools – Fields Ribbon

Table Tools – Table Ribbon

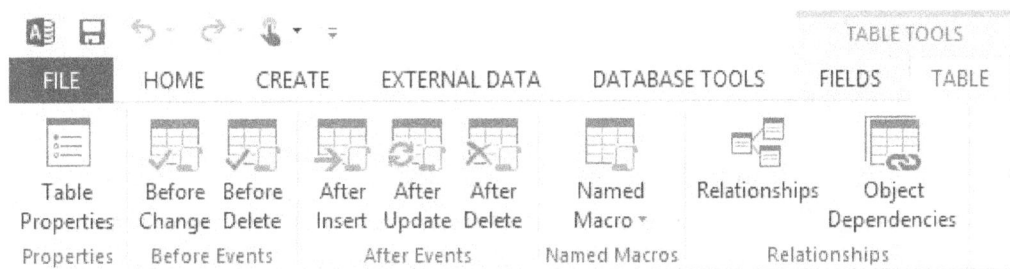

To sort and filter records in datasheet view, use the following procedure.

Step 1: Each column has an arrow with options for sorting and filtering your data. Select the arrow to view the options.

Step 2: To sort the data, select one of the sorting options. To filter the data, check one of the boxes, which are based on the type of data stored in that field. Let's try viewing just the records for students in the Math curriculum. You can choose Select All to clear all the boxes or you can choose more than one box to customize your filter.

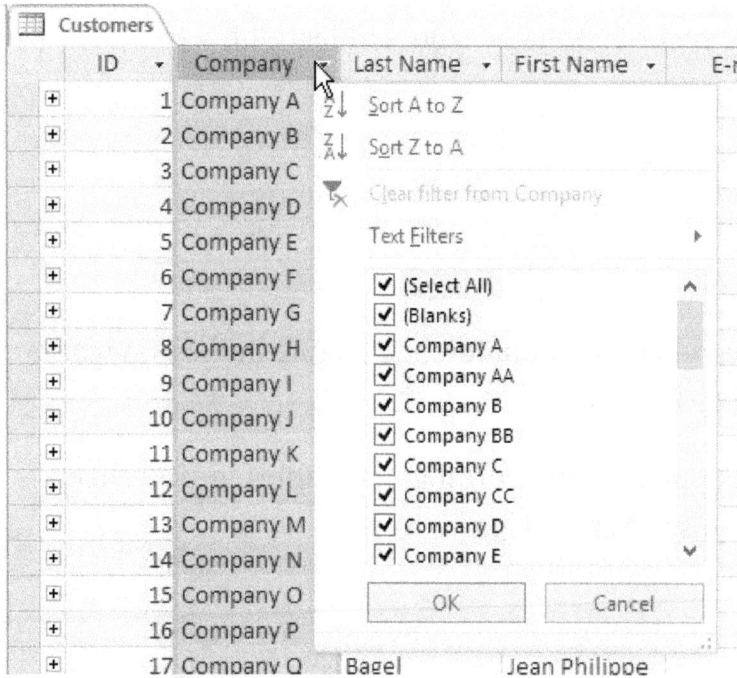

If a sort has been applied, the column has a thin arrow pointing in the direction of the sort next to the arrow for sorting and filtering options. If a filter has been applied, the column has a funnel in the column header. Click on the column header options to change the sort or remove the filter.

About Keys, Data Types and Fields

This lesson looks at three aspects of your database a little more closely to make sure that you understand the concepts before we start working with our tables.

Primary Key is required in each table. There must be one Primary Key and only one. The entry for each record in the primary key differentiates the record from the other records in the table.

Each field has an indicator of what type of data it contains. Each data type has different capabilities or restrictions that you will learn more about as we get deeper into database design. For now, here are the different types of data:

- Short Text – Numbers or letters, with a limit of 255 characters

- Long Text – Text that is too long to be stored in text fields

- Number – Digits only

- Date / Time – A valid date or time

- Currency – Same as number, but with decimal places and a currency symbol added

- AutoNumber – A unique sequential number, such as used for the primary ID

- Yes / No – Accepts yes / no; true/false; on/off

- OLE Object – Any object that can be linked or embedded in a table

- Hyperlink – A path to an object, file or Web site

- Attachment - Any supported type of file, including pictures, charts, text files, and so on

- Lookup Wizard – Creates a drop-down list from existing data or data you enter

You can change the field name from Datasheet View. Just click on the Column header and enter new text.

	Company	Last Name	First Name	E-mail Address	Job Title	Business Phi
1	Company A	Bedecs	Anna		Owner	(123)555-0100
2	Company B	Gratacos Solso	Antonio		Owner	(123)555-0100
3	Company C	Axen	Thomas		Purchasing Representati	(123)555-0100
4	Company D	Lee	Christina		Purchasing Manager	(123)555-0100

Searching for Records

The option to search for a specific record is at the bottom of the datasheet.

To use the Search field. This example uses the Inventory Transactions table in the sample database. The Search box is at the bottom of the screen, use the following procedure.

Step 1: Begin entering the text for which you want to search in the Search box.

Access immediately highlights the first occurrence of the text or numbers that appear in the table.

Step 2: Press Enter to move to the next instance of the Search text.

Deleting Records

You will use the tools on the Home Ribbon or the context menu to delete records.

To delete a record, use the following procedure.

Step 1: Point your mouse to the blue column at the left of the record you want to delete.

You can select the Delete command from the Records menu on the Home Ribbon or right-click and select Delete Record from the context menu.

Step 2: Select Delete Record.

Access may display a contextual warning message. Review the message and click OK or Help.

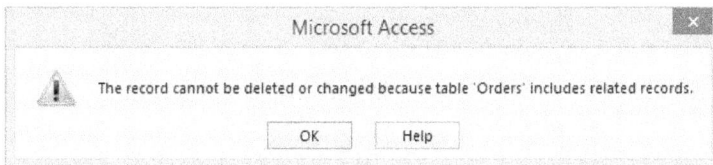

Step 3: Select Yes.

If a field from the record you selected is used in a related table, Access will not allow you to delete the record. It displays a warning message. A sample is illustrated below.

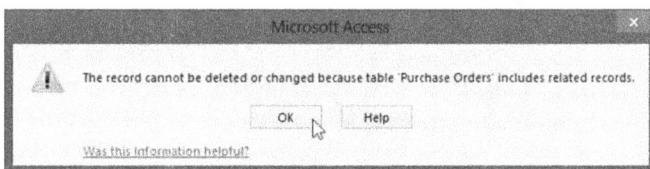

Select OK to close the warning.

If you need to delete the record, you will need to change the data in the related table, so that it no longer references this record.

Chapter 4: Performing Basic Table Tasks

This chapter focuses on tables. You will learn how to enter and edit data in the Datasheet view of a table. Next, we will look at using the clipboard to cut, copy, and paste data to and from different fields. We will also look at how to format text in the table and find and replace data in your tables. Finally, you will learn about saving your table.

Entering and Editing Data

In this lesson, we will look at entering and editing data, which works much the same as it did in the app.

To enter data using a variety of data types, use the following procedure.

In the example Products table, the first field is a multi-answer lookup column. The options appear as checkboxes.

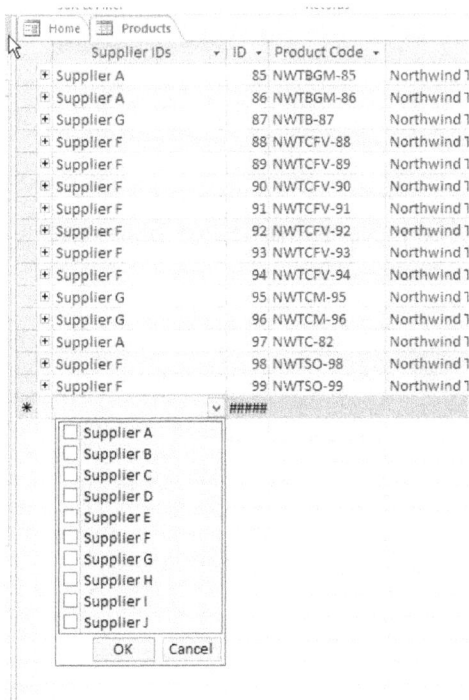

Step 1: Check one or more boxes to enter the data for this field. Select OK.

Step 2: Press Enter or Tab to move to the next field.

In a new record, the Primary Key field (in this example, labeled ID) says New at first, until you start entering information. Then Access enters the next number.

Step 3: Set the formatting elements for each field.

The following fields are TEXT fields. Enter text to complete the information.
- Product Code
- Product Name
- Description
- Quantity per Unit

The following fields are CURRENCY fields. Enter a number, with a decimal for cents, if applicable.
- Standard Cost
- List Price

The following fields are NUMBER fields. Enter a number.
- Reorder Level
- Target Level
- Minimum Reorder Quantity

The Discontinue field is a YES/NO field. Check the box, if applicable.

The Category field is a Lookup column (drop down list). Select an item from the list.

Home | Products

Reorder Le ▾	Target Level ▾	Quantity Per Unit ▾	Discontinue ▾	Minimum Reorder Quantity ▾	Category ▾	ⓤ	Click to Add ▾
10	20	3 boxes	☐	5	Baked Goods & Mixes	ⓤ(0)	
10	20	4 boxes	☐	5	Baked Goods & Mixes	ⓤ(0)	
20	50	100 count per box	☐		Beverages	ⓤ(0)	
10	40	15.25 OZ	☐		Canned Fruit & Vegetables	ⓤ(0)	
10	40	15.25 OZ	☐		Canned Fruit & Vegetables	ⓤ(0)	
10	40	15.25 OZ	☐		Canned Fruit & Vegetables	ⓤ(0)	
10	40	15.25 OZ	☐		Canned Fruit & Vegetables	ⓤ(0)	
10	40	14.5 OZ	☐		Canned Fruit & Vegetables	ⓤ(0)	
10	40	14.5 OZ	☐		Canned Fruit & Vegetables	ⓤ(0)	
30	50	5 oz	☐		Canned Meat	ⓤ(0)	
30	50	5 oz	☐		Canned Meat	ⓤ(0)	
50	200		☐		Cereal	ⓤ(0)	
100	200		☐		Soups	ⓤ(0)	
100	200		☐		Soups	ⓤ(0)	
			☐			ⓤ(0)	
			☐			ⓤ(0)	

Baked Goods & Mixes
Beverages
Candy
Canned Fruit & Vegetabl
Canned Meat
Cereal
Chips
Snacks
Condiments
Dairy Products
Dried Fruit & Nuts
Grains
Jams
Preserves
Oil
Pasta

The final field allows attachments.

To edit a record. Let's imagine that one of the Northwind customers got a promotion, with a new title and contact information. This example uses the Customers table from the sample database.

Step 1: Highlight the information you want to change.

Home | Products | Employees

	ID ▾	Company ▾	Last Name ▾	First Name ▾	E-mail Address ▾	Job Title ▾	Business Ph ▾	Home Phon ▾	M
⊞	1	Northwind Tra	Freehafer	Nancy	nancy@northwindtraders	Sales Representative	(123)555-0100	(123)555-0102	
⊞	2	Northwind Tra	Cencini	Andrew	andrew@northwindtrade	Vice President, Sales	(123)555-0100	(123)555-0102	
⊞	3	Northwind Tra	Kotas	Jan	jan@northwindtraders.cc	Sales Representative	(123)555-0100	(123)555-0102	
⊞	4	Northwind Tra	Sergienko	Mariya	mariya@northwindtrade	Sales Representative	(123)555-0100	(123)555-0102	
⊞	5	Northwind Tra	Thorpe	Steven	steven@northwindtrade	Sales Manager	(123)555-0100	(123)555-0102	
⊞	6	Northwind Tra	Neipper	Michael	michael@northwindtrad	Sales Representative	(123)555-0100	(123)555-0102	
⊞	7	Northwind Tra	Zare	Robert	robert@northwindtrader	Sales Representative	(123)555-0100	(123)555-0102	
⊞	8	Northwind Tra	Giussani	Laura	laura@northwindtraders	Sales Coordinator	(123)555-0100	(123)555-0102	
⊞	9	Northwind Tra	Hellung-Larser	Anne	anne@northwindtraders	Sales Representative	(123)555-0100	(123)555-0102	
✱	(New)								

Step 2: Enter the new information. After you have entered the new information, close the table. There is no need to save – Access saves the new information in the table automatically.

Using the Clipboard

Just as with other applications, you can use the Clipboard to cut, copy, and paste information from one place to another. We will practice these tasks in Datasheet view.

To cut and paste text, use the following procedure.

Step 1: Highlight the text you want to cut.

Step 2: Select Cut from the Home tab on the Ribbon.

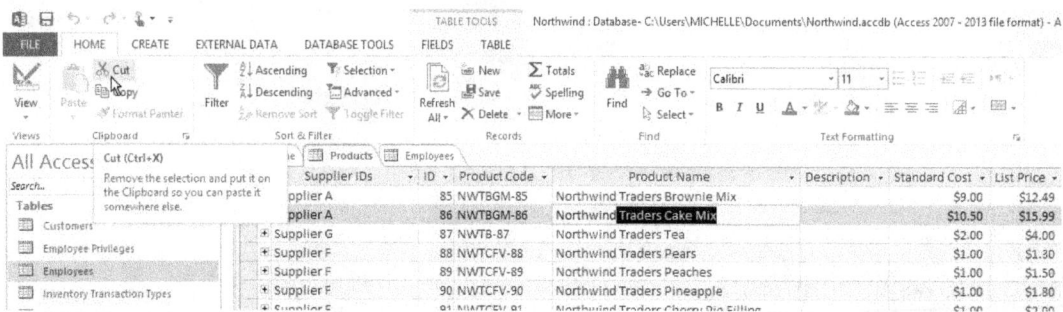

Step 3: Move the cursor to the new location.

Step 4: Right click the mouse and select Paste from the context menu.

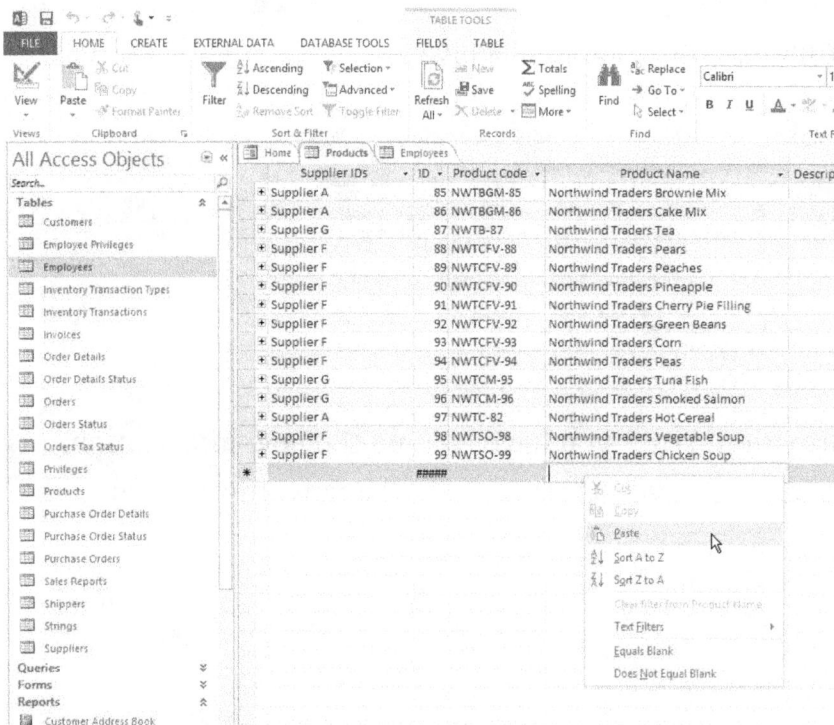

To copy and paste text using the keyboard shortcuts, use the following procedure.

Step 1: Highlight the text you want to cut and press the Control key and the C key at the same time.

Step 2: Move the cursor to the new location.

Step 3: Press the Control key and the V key at the same time.

To open the Clipboard Task pane, select the icon next to Clipboard on the Home tab of the Ribbon.

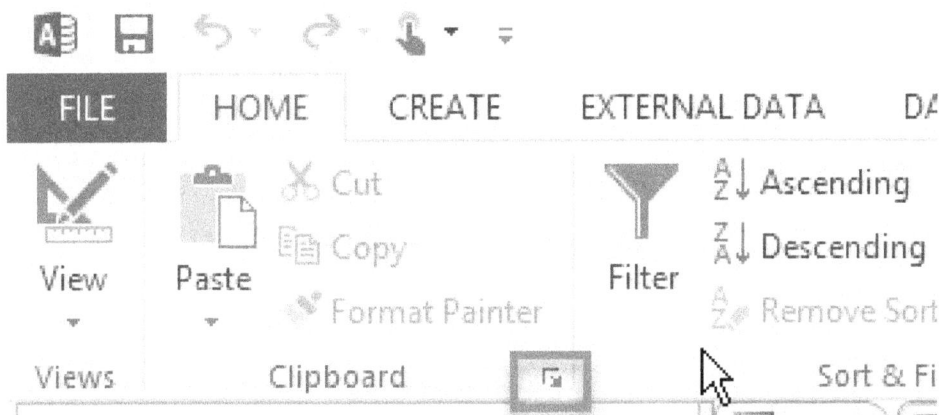

The Clipboard pane opens, displaying any items you have cut or copied in this Word 2016 session (or the 24 most recent). A sample is illustrated below.

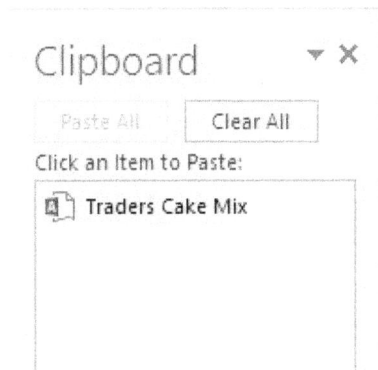

To paste using the Office Clipboard Task pane, use the following procedure.

Step 1: Place the cursor where you want to paste text from the clipboard.

Step 2: Click the item in the Clipboard task pane that you want to paste.

You can copy and paste data from another program like Excel or Word into an Access table. This works best if the data is separated into columns. If the data is in a word processing program, such as Word, either use tags to separate the columns or convert the columns into a table format before copying.

Step 1: If the data needs editing, such as separating full names into first and last names, do that first in the source program.

Step 2: Open the source and copy (Ctrl + C) the data.

Step 3: Open the Access table where you want to add the data in Datasheet view and paste it (Ctrl + V).

Step 4: Double-click each column heading and type a meaningful name.

Step 5: Click File, Save and give your new table a name.

Note that Access sets the data type of each field based on the information you paste into the first row of each column, so make sure that the information in the following rows match the first row.

Formatting Text

There are several options for formatting your text in the Datasheet.

To change the font face and size using the Home Ribbon tools, use the following procedure.

Step 1: Select the text you want to change.

Step 2: Select the arrow next to the current font name to display the list of available fonts.

Step 3: Use the scroll bar or the down arrow to scroll down the list of fonts.

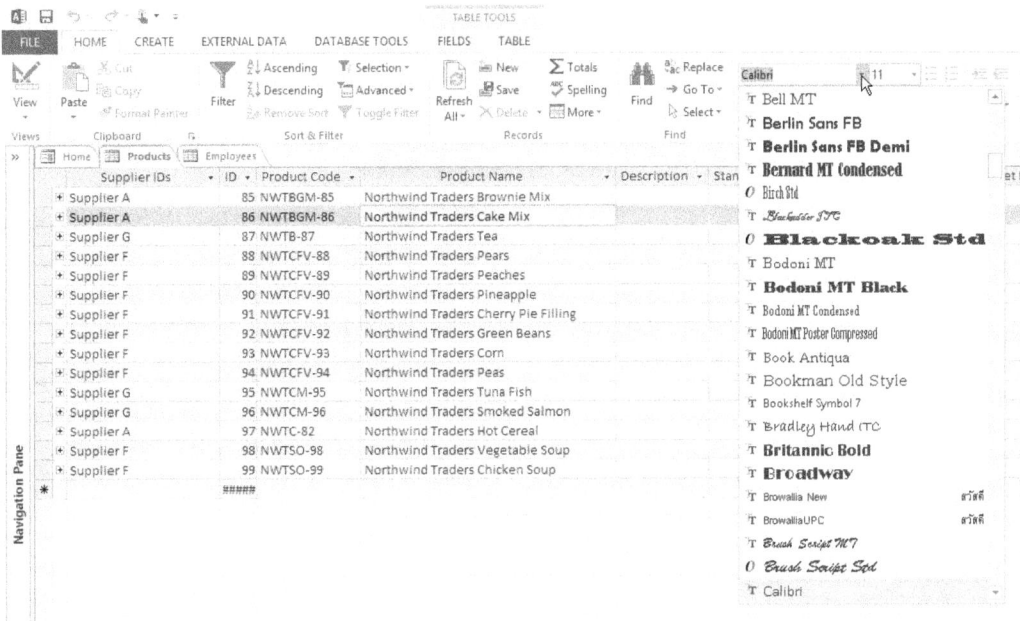

Step 4: Select the desired font to change the font of text.

Step 5: With the text still selected, select the arrow next to the current font size to see a list of common font sizes.

Step 6: Use the scroll bar or the down arrow key to scroll to the size you want and select it. You can also highlight the current font size and type in a new number to indicate the font size you want.

To select a color for their fonts from the gallery, use the following procedure.

Step 1: Select the text you want to change.

Step 2: Select the arrow next to the Font Color tool on the Home Ribbon to display the gallery. Or select the same tool from the context menu (appears when you select text or by right-clicking).

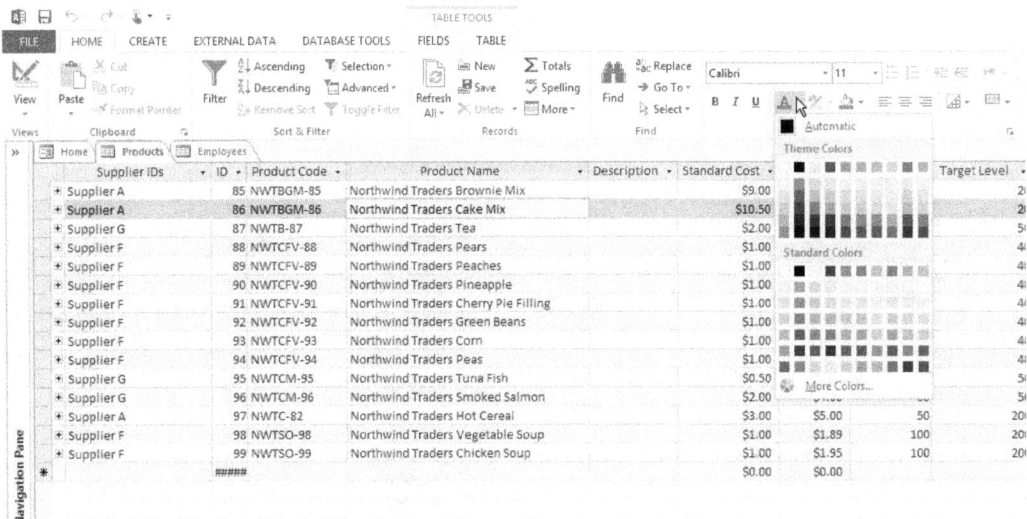

Step 3: Select the color to change the font color.

To review the Colors dialog box, use the following procedure.

Step 1: Select the text you want to change.

Step 2: Select the arrow next to the Font Color tool on the Home Ribbon to display the gallery. Or select the same tool from the context menu (appears when you select text or by right-clicking).

Step 3: Select More Colors to open the Colors dialog box.

In the Standard Colors dialog box, simply click on the color and select OK to use that color.

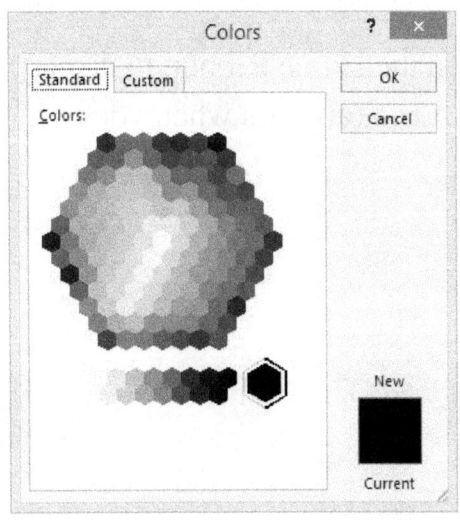

In the Custom Colors dialog box, you can click on the color, or you can enter the red, green, and blue values to get a precise color. When you have the color you want, select OK.

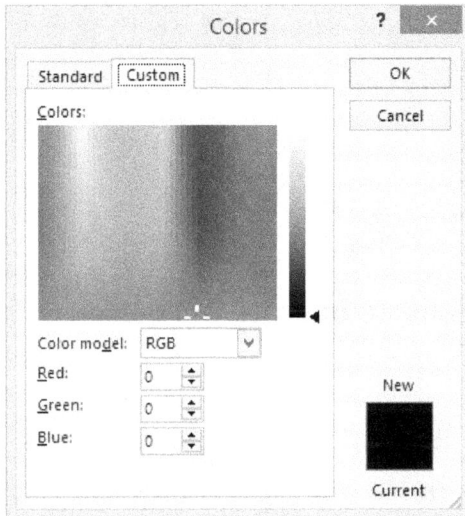

Below are the tools used to add Bold, Italics, and Underline font enhancements.

B *I* U

Finding and Replacing Text

The Find and Replace feature is another one that you may have used in other applications. It is a nice way to make global changes to your Datasheet.

To find and replace one instance of an item at a time, use the following procedure.

Step 1: Select Replace from the Find group on the Home tab of the Ribbon.

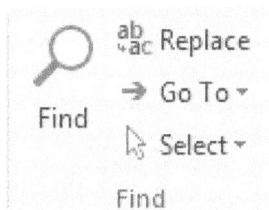

Step 2: In the Find and Replace dialog box, enter the exact text you want to find in the Find what field.

Step 3: Enter the replacement text in the Replace with field.

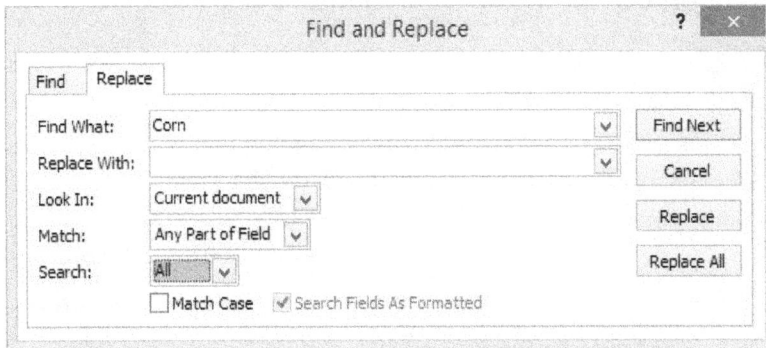

Step 4: Select an option from the Look In drop down list to indicate whether Access should look in the current field or the current document.

Step 5: Select an option from the Match drop down list to indicate whether Access should look for any part of the field, the whole field, or the start of the field.

Step 6: Select an option from the Search drop down list to indicate whether Access should search up, down, or all (the whole datasheet).

Step 7: Check the Match Case box if applicable.

Step 8: Check the Search Fields As Formatted box if applicable.

Step 9: Select Find next to find the next instance of the item.

Access highlights the record where it finds the search term. It also highlights the item from the Find and Replace dialog box.

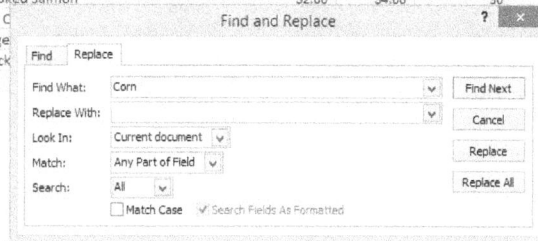

Supplier IDs	ID	Product Code	Product Name	Description	Standard Cost	List Price	Reorder Le	Target Level
lier A	85	NWTBGM-85	Northwind Traders Brownie Mix		$9.00	$12.49	10	
lier A	86	NWTBGM-86	Northwind Traders Cake Mix		$10.50	$15.99	10	
lier G	87	NWTB-87	Northwind Traders Tea		$2.00	$4.00	20	
lier F	88	NWTCFV-88	Northwind Traders Pears		$1.00	$1.30	10	
lier F	89	NWTCFV-89	Northwind Traders Peaches		$1.00	$1.50	10	
lier F	90	NWTCFV-90	Northwind Traders Pineapple		$1.00	$1.80	10	
lier F	91	NWTCFV-91	Northwind Traders Cherry Pie Filling		$1.00	$2.00	10	
lier F	92	NWTCFV-92	Northwind Traders Green Beans		$1.00	$1.20	10	
lier F	93	NWTCFV-93	Northwind Traders Corn		$1.00	$1.20	10	
lier F	94	NWTCFV-94	Northwind Traders Peas		$1.00	$1.50	10	
lier G	95	NWTCM-95	Northwind Traders Tuna Fish		$0.50	$2.00	30	
lier G	96	NWTCM-96	Northwind Traders Smoked Salmon		$2.00	$4.00	30	
lier A	97	NWTC-82	Northwind Traders Hot C					2
lier F	98	NWTSO-98	Northwind Traders Vege					2
lier F	99	NWTSO-99	Northwind Traders Chick					2

Step 10: When Access highlights the item, select Replace to delete the "find" item and paste the "replace" item.

Step 11: Select Cancel when you have finished.

To Replace all instances of an item, use the following procedure.

Step 1: Open the Find and Replace dialog box by selecting Replace from the Home Ribbon.

Step 2: Enter the exact text you want to find in the Find what field.

Step 3: Enter the replacement text in the Replace with field.

Step 4: Select Replace All.

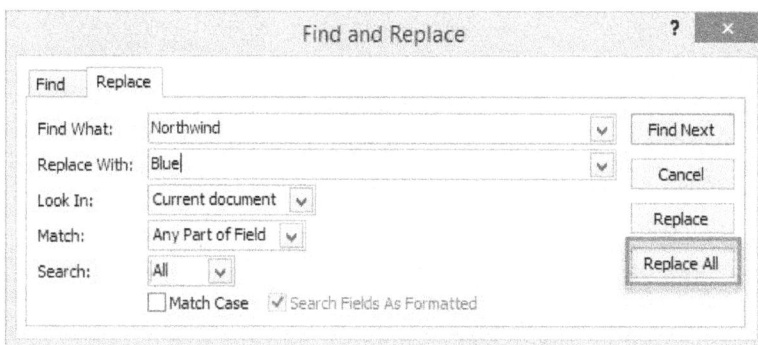

Access replaces all instances of the item in the current datasheet. Access displays a warning message.

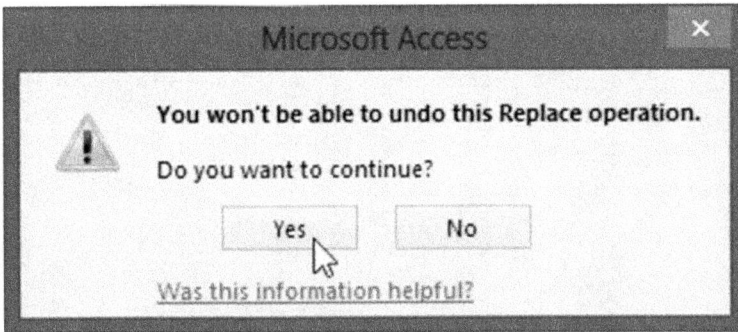

Step 5: Select Cancel when you have finished.

Chapter 5: Working with Fields

Now we will start getting more into database design. First, we will look at adding a field by entering data. Then we will look at adding a specific type of field. You will learn how to change the field name, caption, or description for the field and how to change the field format. Finally, we will look at deleting a field.

Adding a Field by Entering Data

You can add a field to your table simply adding the type of data you want that field to include.

To add a field by entering data, use the following procedure.

Step 1: As you are entering data in a table, or even if you are editing a previous record, click on the column labeled "Click to Add". You cannot tab from the previous field; you will need to click on that column.

	State/Provir ▾	ZIP/Postal C ▾	Country/Reg ▾	Web Page ▾	Notes ▾	🖉	Click to Add ▾
⊞ WA	99999	USA	http://northwi		🖉(0)	
⊞ WA	99999	USA	http://northwi	Joined the com	🖉(0)	
⊞ WA	99999	USA	http://northwi	Was hired as a	🖉(0)	
⊞ WA	99999	USA	http://northwi		🖉(0)	
⊞ WA	99999	USA	http://northwi	Joined the com	🖉(0)	
⊞ WA	99999	USA	http://northwi	Fluent in Japan	🖉(0)	
⊞ WA	99999	USA	http://northwi		🖉(0)	
⊞ WA	99999	USA	http://northwi	Reads and writ	🖉(0)	
⊞ WA	99999	USA	http://northwi	Fluent in Frenc	🖉(0)	
✱					🖉(0)	

Step 2: Enter data in the column. When you press Tab, or Enter after the first entry, another new column displays, where you can continue creating new fields.

	Address ▾	City ▾	State/Provir ▾	ZIP/Postal C ▾	Country/Reg ▾	Web Page ▾	Notes ▾	🖉	Click to Add ▾
⊞ 123 1st Avenue	Seattle	WA	99999	USA	http://northwi		🖉(0)	
⊞ 123 2nd Avenu	Bellevue	WA	99999	USA	http://northwi	Joined the com	🖉(0)	
⊞ 123 3rd Avenue	Redmond	WA	99999	USA	http://northwi	Was hired as a	🖉(0)	
⊞ 123 4th Avenue	Kirkland	WA	99999	USA	http://northwi		🖉(0)	
⊞ 123 5th Avenue	Seattle	WA	99999	USA	http://northwi	Joined the com	🖉(0)	
⊞ 123 6th Avenue	Redmond	WA	99999	USA	http://northwi	Fluent in Japan	🖉(0)	
⊞ 123 7th Avenue	Seattle	WA	99999	USA	http://northwi		🖉(0)	
⊞ 123 8th Avenue	Redmond	WA	99999	USA	http://northwi	Reads and writ	🖉(0)	
⊞ 123 9th Avenue	Seattle	WA	99999	USA	http://northwi	Fluent in Frenc	🖉(0)	
✱							🖉(0)	

Adding a Specific Type of Field

While Access usually gets the data type correct when you enter data to create a field as in the previous lesson, sometimes you may want to make sure that it is the right data type before you enter your data.

To add a specific type of field, use the following procedure.

Step 1: Select the arrow next to the Click to Add column header.

Step 2: Select the type of field from the drop-down list.

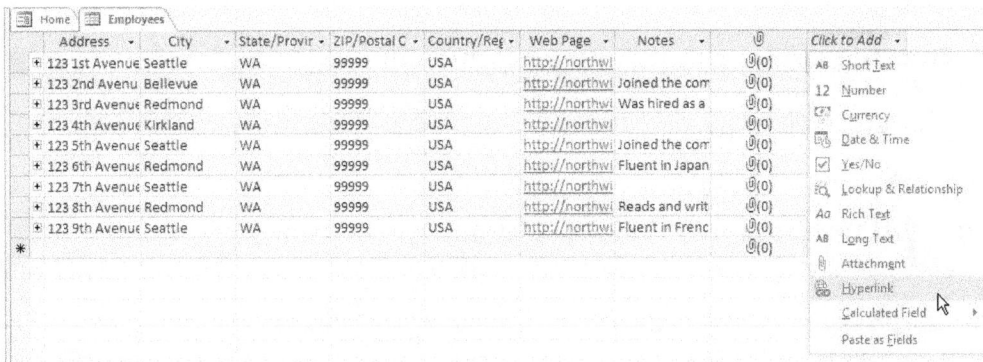

Step 3: Enter your data.

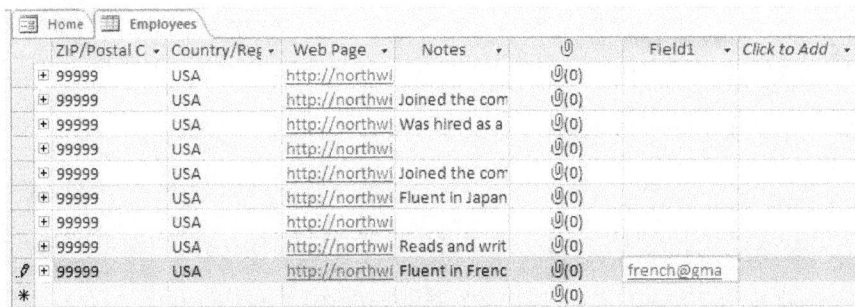

For more types of fields, you can use the Add & Delete group on the Fields tab of the Table Tools Ribbon. Select the type of field to add it. Select More Fields for more choices on types of fields.

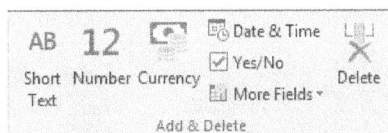

Changing Field Name, Caption, or Description

When you add fields either by entering data or by choosing the data type first, Access labels the field with Field1 and subsequent numbers. This lesson looks at how to change that, as well as the field caption or description.

To change the field name, caption, and description in the Enter Field Properties dialog box, use the following procedure.

Step 1: Make sure the Fields tab on the Table Tools Ribbon is showing.

Step 2: Select Name & Caption.

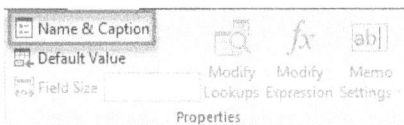

Step 3: In the Name box, enter the field name.

Step 4: Enter the Caption text. Captions are used as the column headers and in other areas of Access or other views.

Step 5: Enter a Description of the field.

Step 6: Select OK.

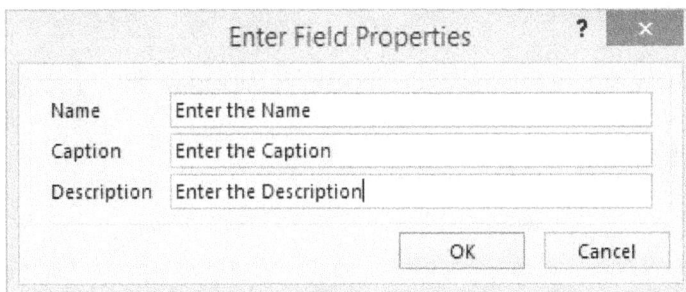

Changing the Data Type

Once you have set up your field, you may want to realize that you do not have the right data type. You can always change it later.

To change the data type, use the following procedure.

Step 1: Highlight the field that you want to change.

Step 2: Make sure the Fields tab is showing on the Table Tools Ribbon.

Step 3: Select a new option from the Data Type drop down list.

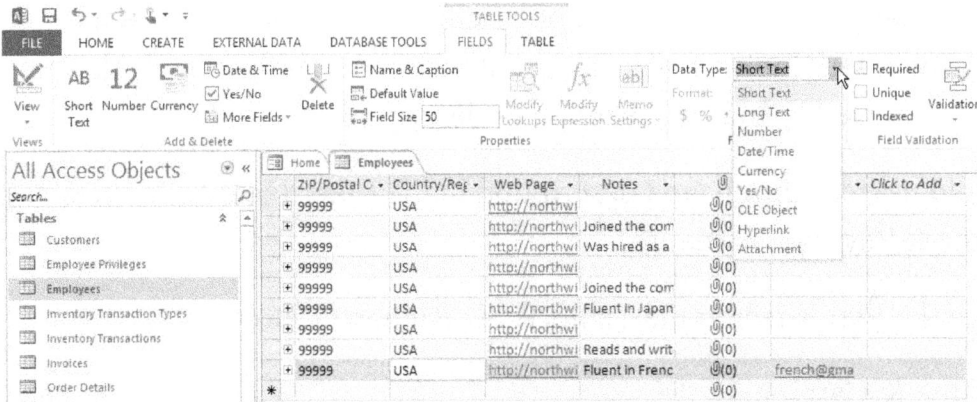

Step 4: Depending on what type of change you made, Access may display a warning message. Select Yes if you understand the risk, and it is the change you meant, or select No to cancel and select a different data type. In this example, we changed from the Hyperlink data type to short text.

Changing Field Format

Some fields include a format, such as numbers and date/time fields. You can change the formatting just like you changed the type of field.

To change the data type, use the following procedure.

Step 1: Highlight the field that you want to change.

Step 2: Make sure the Fields tab is showing on the Table Tools Ribbon.

Step 3: Select a new option from the Format drop down list.

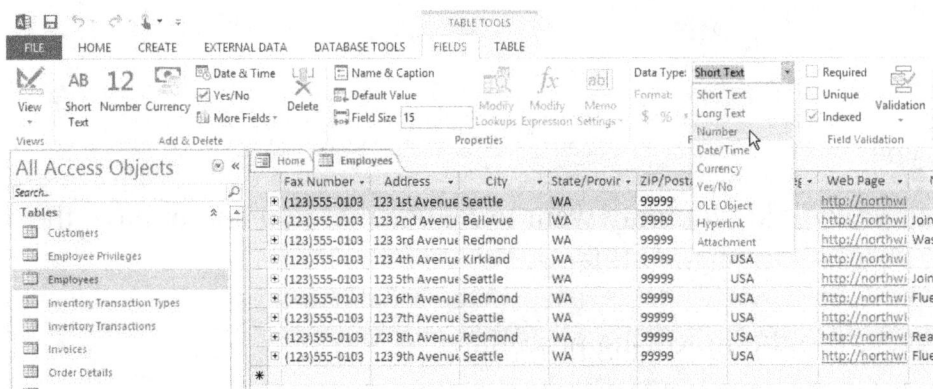

Practice adding different types of fields to see the options available for formatting.

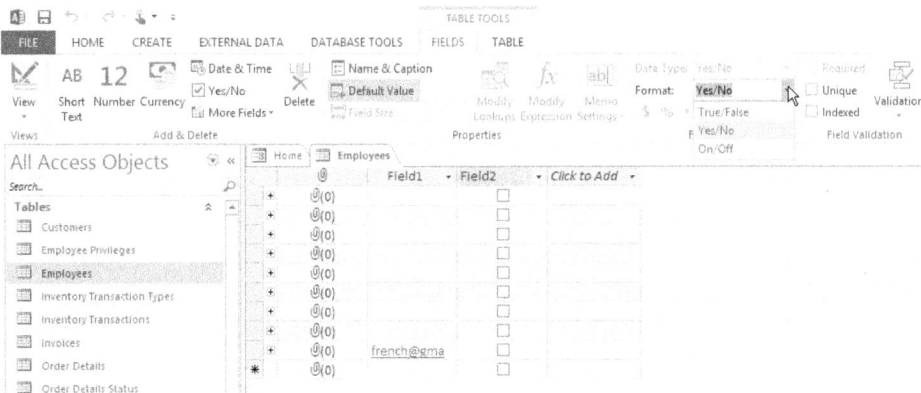

Deleting a Field

If your fields are not related to another table, you can delete them if you realize that you do not need the field.

To delete a field, use the following procedure. In this example, we will try to delete a field with a table relationship.

Step 1: Click the column header for the field you want to delete.

Step 2: Make sure the Fields tab on the Table Tools Ribbon is showing.

Step 3: Select Delete.

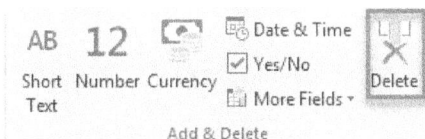

Step 4: If you selected a field that is part of a table relationship, Access displays a warning that it cannot delete that field.

Now let's practice a field without a relationship, use the following procedure.

Step 1: Click the column header for the field you want to delete.

Step 2: Right-click and select Delete field from the context menu.

Step 3: Access displays a warning to make sure that you want to delete the data. There is no undo. Select Yes to continue.

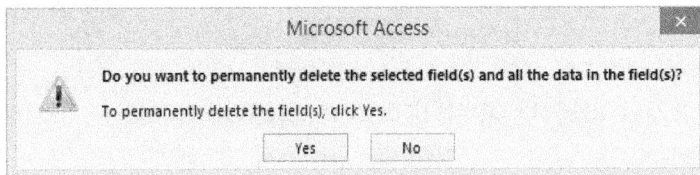

Chapter 6: Working with Table Relationships

This chapter will teach students about the different types of relationships, how to view relationships, and how to edit relationships. Students will also learn about referential integrity and how to establish it.

Types of Relationships

Table relationships are the basis of a relational database like Access.

Relationships are a means to join data to different tables, while avoiding redundancy in the tables. Therefore, you can divide your data into different tables—using it only once—and then add it into other tables by establishing relationships.

There are three types of relationships:

A One-to-Many relationship is where each entry on table 1 can have a relationship with multiple entries on table 2, but not vice versa. For an example, let's consider a database that tracks orders and customers. A customer can place many orders, which would be represented by a one-to-many relationship in the Customers table. The Primary Key (or Customer ID) field in the Customers table is also a field in the Orders table (though not the primary key). This allows Access to locate the correct customer for each order.

A Many-to-Many relationship is where each entry on table 1 can have a relationship with multiple entries on table 2 and each entry on table 2 can have a relationship with multiple entries on table 1. For an example, let's consider a database that tracks both products and orders. One order can include multiple products, as products can be referenced by multiple items in the Orders table. Consider both sides of the relationship when using a many-to-many relationship.

A One-to-one relationship is where each entry on table 1 can only have a relationship with one entry on table 2 and each entry on table 2 can only have a relationship with one entry on table 1. These types of relationships are rare because the information is usually stored in the same table. Reasons to use a one-to-one relationship might be to split a table with many fields, to isolate part of a table for security reasons, or to store information that applies only to a subset of the main table. Both tables in this relationship must share a common field.

One big reason for creating table relationships: creating relationships provides a foundation for establishing referential integrity, which will be discussed in a later section.

Viewing Relationships

Access displays table relationships in a visual way to make it easy to see which fields are related.

To view the relationships in a database, use the following procedure.

Step 1: Select the Database Tools tab from the Ribbon.

Step 2: Select Relationships.

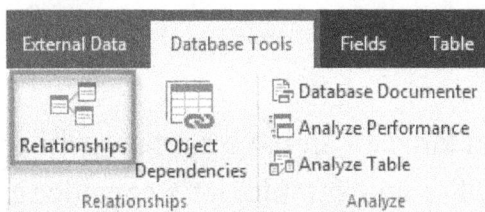

Step 3: Depending on what kind of relationships are available in your database, you can select Direct Relationships or All Relationships

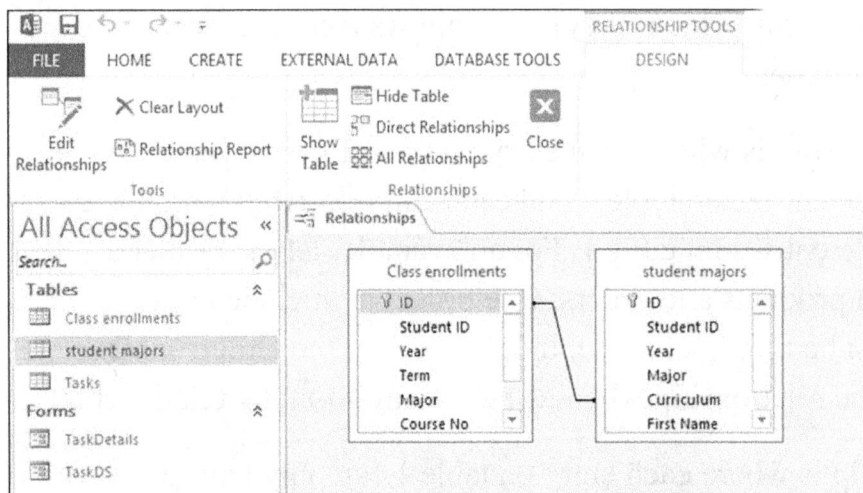

Step 4: Select Close to close the Relationships window.

The Northwind database has a much more complicated Relationships window. The icons indicate whether the relationship is one-to-many or many-to-one in a relationship with referential integrity.

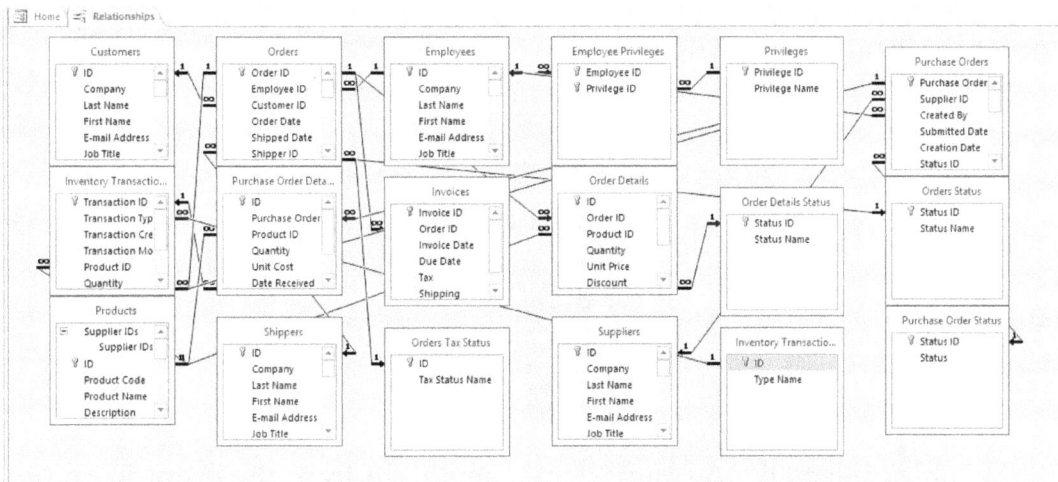

Editing Relationships

You can edit a relationship in Access 2016.

To open the Edit Relationships dialog box, use the following procedure.

Step 1: Double-click on a relationship indicator in the Relationships window or click on the relationship indicator and select Edit Relationships from the Design tab of the Relationship Tools Ribbon.

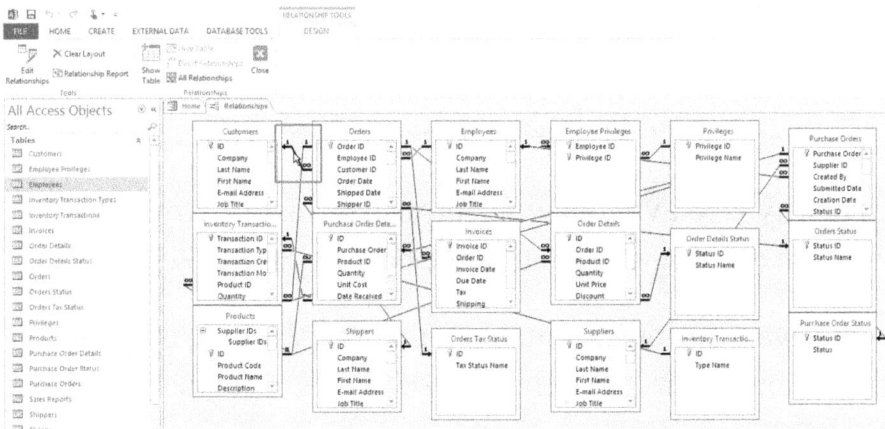

Step 2: In the Edit Relationships dialog box, you can select a new table from either side of the relationship.

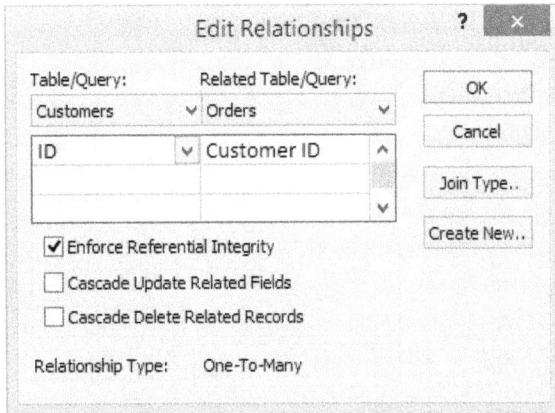

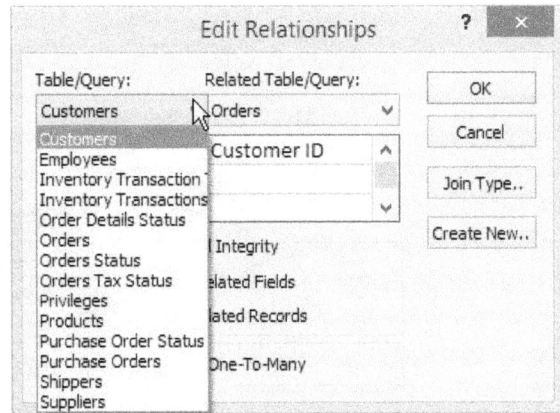

Step 3: With the tables selected, you can select a new field from the drop-down list on either side.

Step 4: Select OK.

About Referential Integrity

Referential integrity is a setting in the relationship to help protect your data.

The goal of referential integrity is to avoid having "orphaned" data. "Orphaned" data can happen when you are deleting or updating the data in your tables.

For example, imagine you have a customer on table 1 that is linked to payments on table 2. If you delete that customer, the payments linked to that customer on table 2 will become "orphaned" data.

Referential integrity prevents this by denying changes that will result in "orphaned" data.

Since you may have a valid need to delete or update data in your tables, you can choose to allow Access to update or delete data that have a relationship with the data you are updating or deleting.

Establishing Referential Integrity

The Referential Integrity options are in the Edit Relationships dialog box.

To enforce referential integrity, use the following procedure.

Step 1: In the Edit Relationships dialog box, check the Enforce Referential Integrity box.

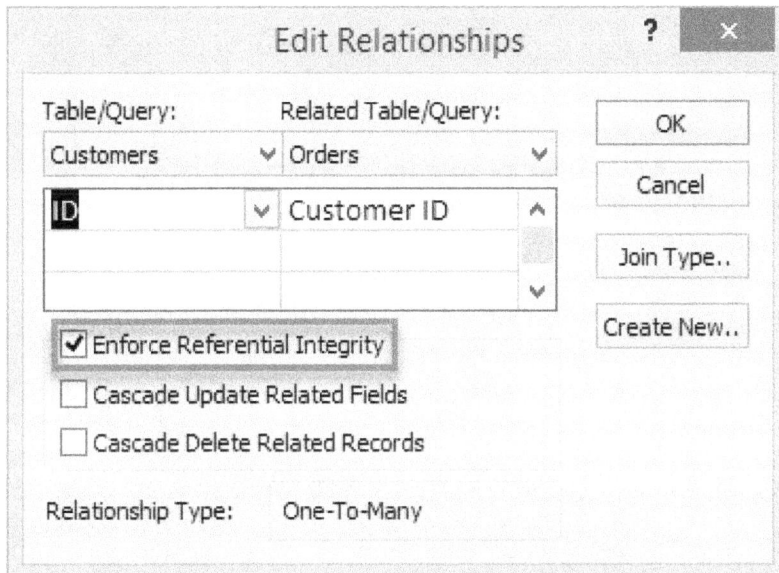

Step 2: The choices to "Cascade update related fields" or "Cascade delete related records" will open. Put a check mark in one or both boxes if you want to allow Access to update or delete data that have a relationship with the data you are updating or deleting.

Step 3: Select OK.

Chapter 7: An Introduction to Queries, Forms, and Reports

This chapter introduces you to some of the other objects in Access. You will learn about the types of queries and learn to create a query with the Query Wizard. Next, we will look at forms and form views. Finally, we will look at reports in Access.

Types of Queries

In this lesson, we will look at the two categories of queries.

Select Queries

Select queries are queries that retrieve data or make calculations. These queries can answer simple questions or combine data from different tables.

Here are some sample select queries:

- Review a subset of data from one table
- Review a subset of data from more than one table
- Ask variations of a question about your data
- Make calculations based on the data, including totals queries
- Summarize and group data, including crosstab queries

Action Queries

Action queries are queries that can add, change, or delete data from your tables.

Here are some sample action queries:

- Create a new table with a make table query
- Add data to an existing table with an append query
- Make automated changes to table data with an update query
- Make automated deletions of table data with a delete query

Creating a Query with the Wizard

The Query wizard makes it easy to create a new query.

To use the Query Wizard to create a simple query, use the following procedure.

Step 1: Select the Create tab from the Ribbon.

Step 2: Select Query Wizard.

Access displays the New Query dialog box. The four choices for queries using the Query Wizard are:

- Simple Query Wizard
- Crosstab Query Wizard
- Find Duplicates Query Wizard
- Find Unmatched Query Wizard

In this example, we will create a simple query to display customer phone numbers.

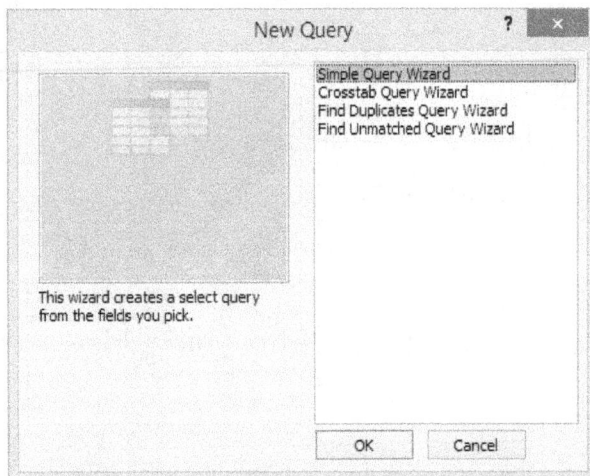

Step 3: Select Simple Query Wizard and select OK.

The first screen in the Wizard allows you to select the table or other query where you want to obtain the data for your query.

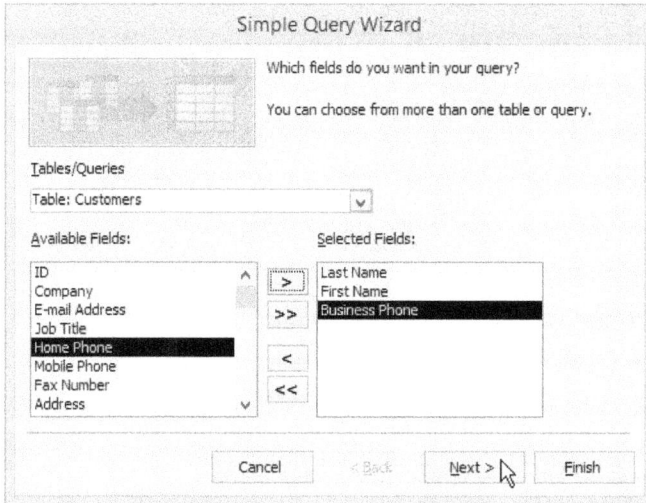

Step 4: You can select more than one table or query for the data you want on your query. If you highlighted a table in the Navigation pane before starting the query wizard, that table is selected. However, you can change it by selecting a new item from the Tables/Queries drop down list.

Step 5: The fields available on the selected table appear in the Available Fields column. Double-click the fields you want on your form, or highlight the field(s) and select the right arrow (or the double right arrow to select all). The items in the Selected Fields column will appear on your query. To remove an item from the Selected Fields column, highlight it and select the left arrow (or the double left arrow to remove all). To add fields from an additional table, return to step 3.

Step 6: When you have finished selecting the fields to appear on your query, select Next.

Step 7: On the next screen, enter a Name for your query.

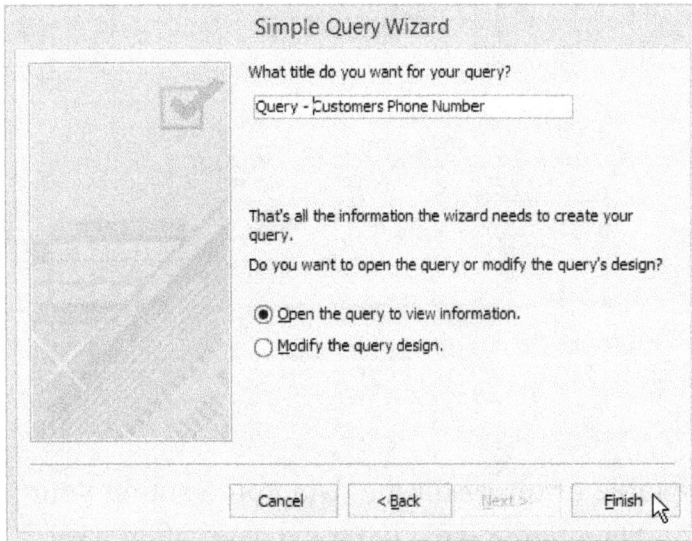

Step 8: Modify the default query name, if desired. Select whether to open the query to view the information, or to modify the query's design. Select Finish.

A sample query is illustrated below.

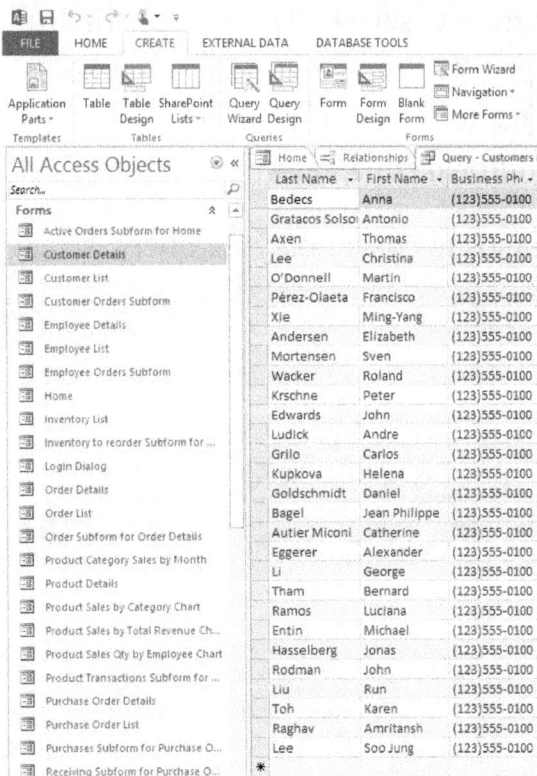

Executing a Query

Once you have created a query, you can execute it at any time. This is helpful if your data changes.

To execute a query, use the following procedure.

Step 1: Find the query that you want to execute in the Navigation pane.

Step 2: Double-click or press Enter.

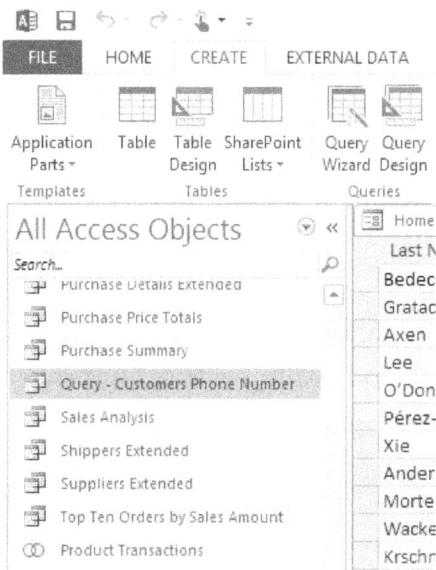

Understanding Forms

Forms are another database object. Forms also work with tables. Forms are a user-friendly way of collecting or displaying the information that is stored in the tables.

To open a form, use the following procedure.

In the Navigation pane showing types of objects, select the double arrows to the right of Forms. Double-click any form to open it. We will look at the Customer Details in this example.

Notice the icons next to the forms.

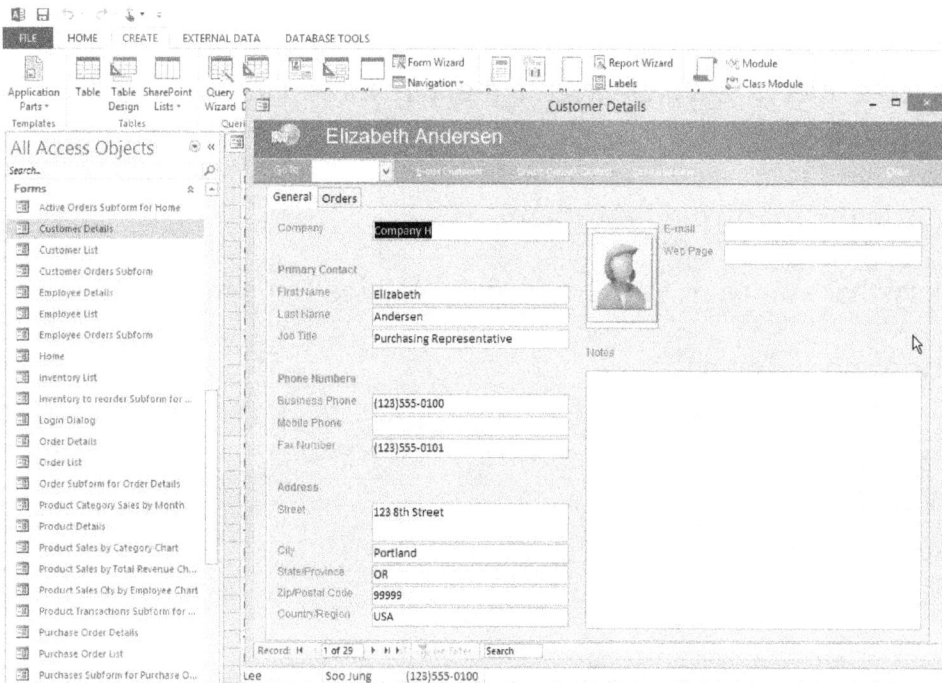

Notice the tabs to open additional information for the selected form. Notice the actions at the top of the form. Notice the navigation options at the bottom to move to the first, previous, next, or last record in the table. You can also perform a search on the data or filter it.

Understanding Reports

Reports take the information you have collected in forms, stored in tables, and manipulated with queries, and they provide a clean way to review or print the data.

To open a report, use the following procedure.

In the Navigation pane showing types of objects, select the double arrows to the right of Reports. Double-click any report to open it. We will look at the Customer Address Book in this example.

Notice the icons next to the reports.

FILE | HOME | CREATE | EXTERNAL DATA | DATABASE TOOLS

Application Parts ~ | Table | Table Design | SharePoint Lists ~ | Query Wizard | Query Design | Form | Form Design | Blank Form | Form Wizard / Navigation ~ / More Forms ~ | Report | Report Design | Blank Report | Report Wizard / Labels | Macro | Module / Class Module / Visual Basic

Templates | Tables | Queries | Forms | Reports | Macros & Code

All Access Objects ⊙ «

Search... 🔎

🗐 Purchase Details Extended	
🗐 Purchase Price Totals	
🗐 Purchase Summary	
🗐 Query - Customers Phone Number	
🗐 Sales Analysis	
🗐 Shippers Extended	
🗐 Suppliers Extended	
🗐 Top Ten Orders by Sales Amount	
⊘ Product Transactions	

Forms ⊻
Reports ⊼

🗐 Customer Address Book	
🗐 Customer Phone Book	
🗐 Employee Address Book	
🗐 Employee Phone Book	
🗐 Invoice	
🗐 Monthly Sales Report	
🗐 Product Category Sales by Month	

Home | Relationships | Query - Customers Phone Number | Customer Address Book

Customer Address Book

Customer Name	Address	City	State
A			
Elizabeth Andersen	123 8th Street	Portland	OR
Catherine Autier Miconi	456 18th Street	Boston	MA
Thomas Axen	123 3rd Street	Los Angelas	CA
B			
Jean Philippe Bagel	456 17th Street	Seattle	WA
Anna Bedecs	123 1st Street	Seattle	WA
E			
John Edwards	123 12th Street	Las Vegas	NV
Alexander Eggerer	789 19th Street	Los Angelas	CA
Michael Entin	789 23th Street	Portland	OR
G			

Chapter 8: Protecting Your Data

To protect the hard work of designing your database and collecting the data, you should plan regular backups. This chapter talks about guidelines and tips for planning backups as well as how to back up a database. You will also learn how to restore a whole database, as well as individual objects.

Planning Backups

You should plan regular backups to protect your data.

If there is a system failure, you will need a backup copy of your desktop database to restore the entire database or to restore an object.

Though backups take up space, they will save you time by avoiding data and design loss. A regular backup schedule is especially important if you have several users updating a database.

Here are some guidelines and tips to help you decide when to back up your database.

For this type of database…	Create backups …
An archive or reference that rarely changes,	Only when the data or design changes.
An active database,	According to a regular schedule that works in your organization.
A database with multiple users,	After a design change. All users of the database will need to close the database so that all changes to the data are saved in the backup.

You should consider making a backup before you run any type of action query, especially if the query will change or delete data.

Backing Up a Database

Now that you have figured out when to make backups, let's look at how to make them.

To make a backup of the database.

Step 1: Select the File tab from the Ribbon to open the Backstage View.

Step 2: Select Save As.

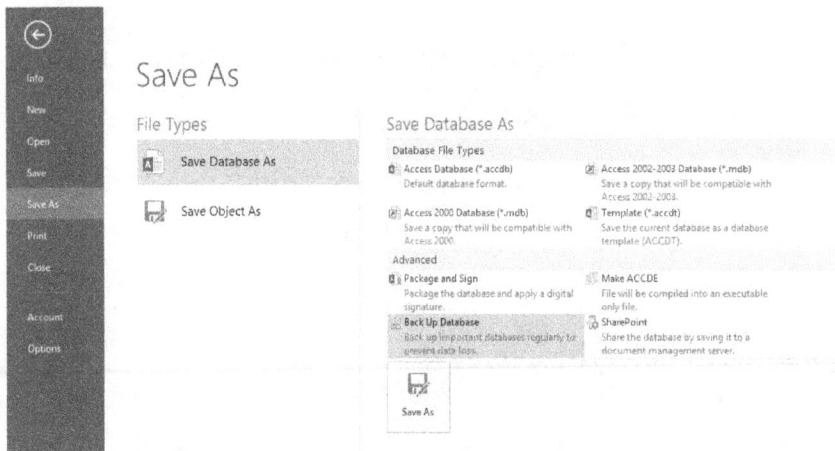

Step 3: Select Back Up Database under the Advanced heading.

Access displays the Save As dialog box.

Step 4: The default file name captures both the name of the original database and the date on which you are making the backup. However, you can change the name and location of the backup as desired.

Step 5: Select Save.

Restoring a Database

If you have created a "known good copy" of your database, such as a backup, you can use it to restore your database if something goes wrong.

To restore a database, use the following procedure.

Step 1: Open File Explorer.

Step 2: Navigate to the location of the backup copy or the known good copy of the database.

Step 3: Copy the backup copy to the location where the damaged or missing database should be replaced.

Step 4: If you are prompted to replace an existing file, do so.

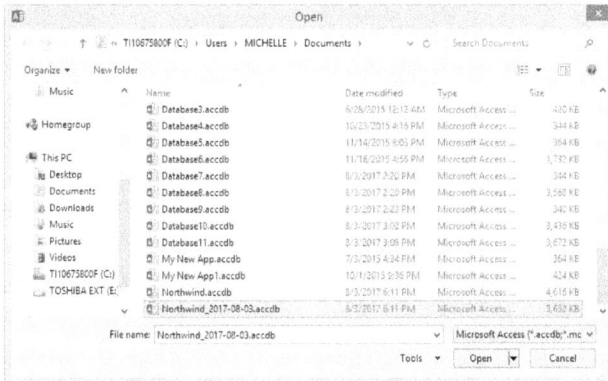

Restoring Objects in a Database

You can also just restore certain objects in a database.

To restore an object to a database from a backup copy, use the following procedure.

Step 1: Open the database to which you want to restore an object.

Step 2: If you want to preserve an object with bad or missing data or that is damaged to compare with the restored version, rename the object. For example, you could rename an object named Tasks with Tasks bad.

Step 3: Delete the object that you want to replace.

Step 4: Select the External Data tab from the Ribbon.

Step 5: Select Access.

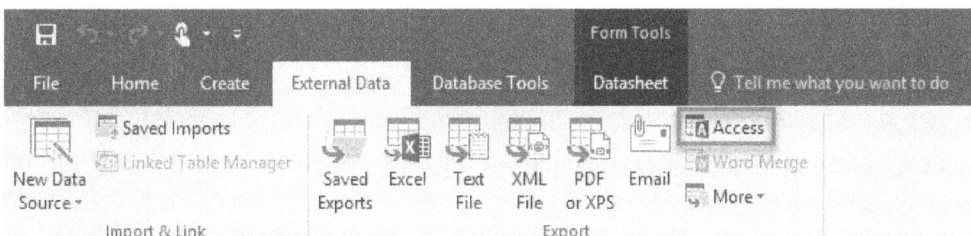

Step 6: In the Get External Data dialog box, select Browse.

Step 7: Navigate to the location of the backup copy of the database. Select Open.

Step 8: Make sure the Import tables… option is selected. Select OK.

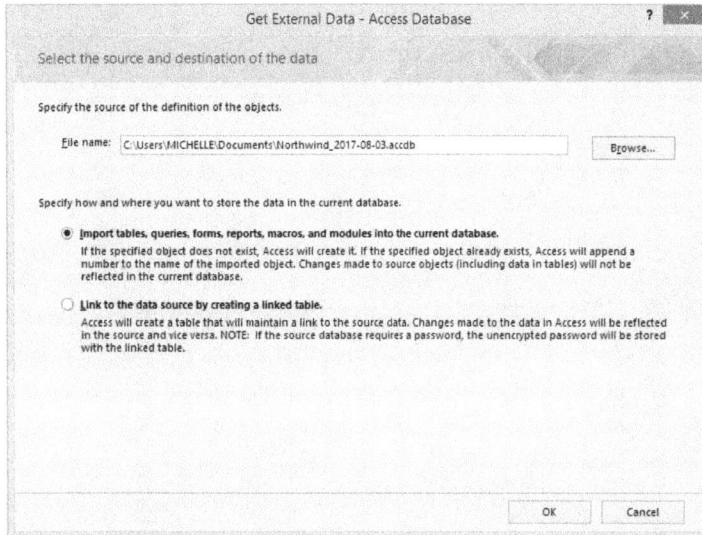

Step 9: In the Import Objects dialog box, select the Tab that corresponds to the type of object that you want to restore. Select the object(s) that you want to restore. Select OK.

Step 10: You can save the import steps if desired. Select Close.

Chapter 9: Working with More Record Tasks

This chapter shows you how to use additional tools for getting your datasheet looking like you want. First, you will learn how to adjust row height and column width in your datasheet. You will also learn how to hide/unhide and freeze/unfreeze fields. Finally, this chapter looks at the connection with Outlook, which allows you to add contacts from Outlook or save a record as an Outlook Contact.

Adjusting Row Height and Width

The Records area of the Home tab also allows you to adjust the selected row height or the selected column width.

To adjust row height, use the following procedure.

Step 1: Select the Home tab from the Ribbon.

Step 2: Select More from the Records menu.

Step 3: Select Row Height.

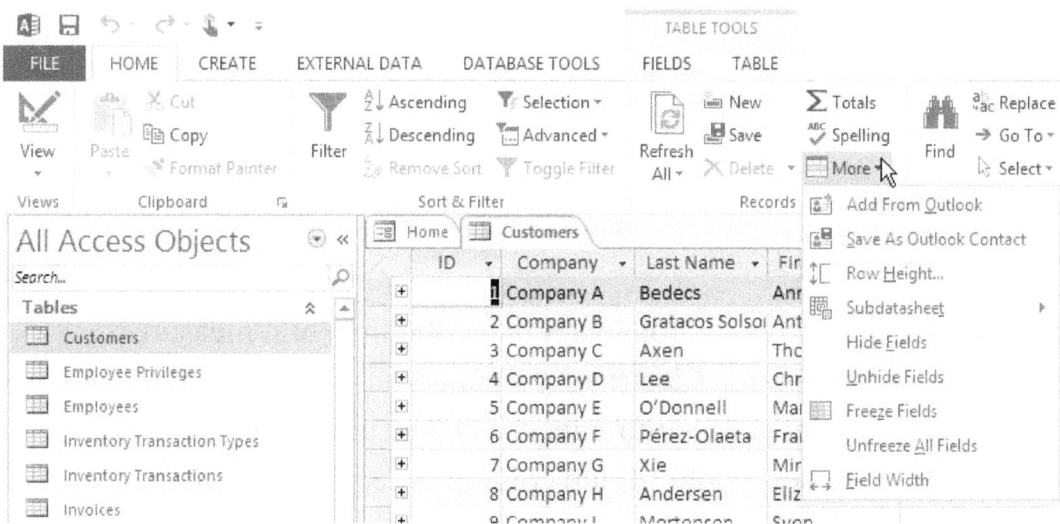

Step 4: In the Row Height dialog box, enter the Row Height measurement in pixels that you want to use for your datasheet row height. Or check the Standard Height box to return to the default measurement.

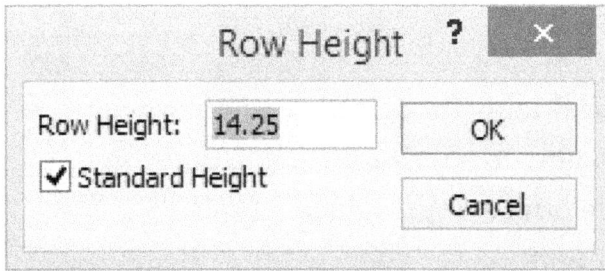

Step 5: Select OK.

To adjust column width, use the following procedure.

Step 1: Select the Home tab from the Ribbon.

Step 2: Select More.

Step 3: Select Field Width.

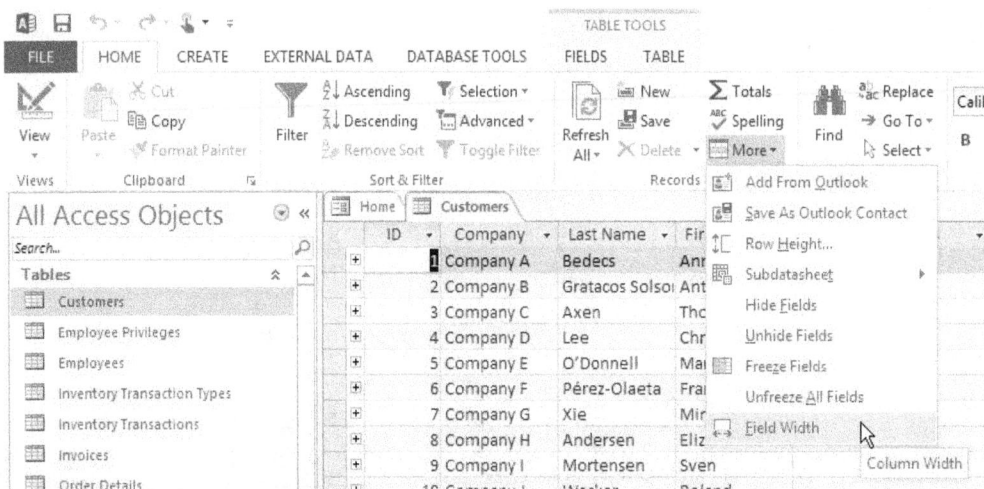

Step 4: In the Column Width dialog box, enter the Column Width measurement in pixels that you want to use for your datasheet column width. Or check the Standard Width box to return to the default measurement. You can also select Best Fit to have Access determine the best width for the column based on the contents.

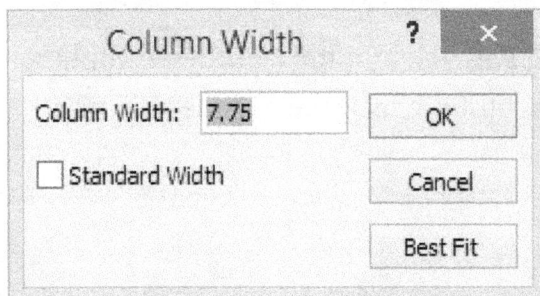

Step 5: Select OK.

Hiding and Un-hiding Fields

You can hide fields that you do not currently need to view in your datasheet.

To hide a field, use the following procedure.

Step 1: Open the table you want to modify in Datasheet view.

Step 2: Select the Home tab from the Ribbon.

Step 3: Select More.

Step 4: Select Hide Fields.

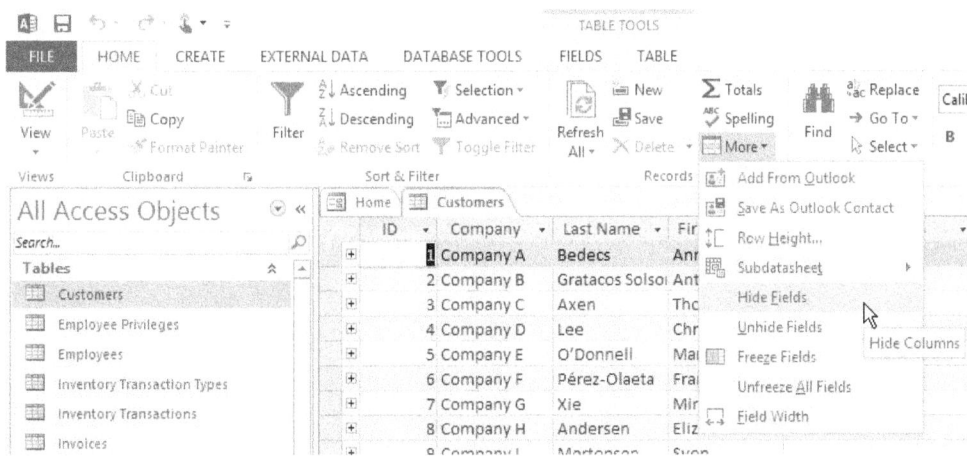

Step 5: The field is no longer shown on the table.

To unhide a field, use the following procedure.

Step 1: Right-click any column header.

Step 2: Select Unhide Fields from the context menu.

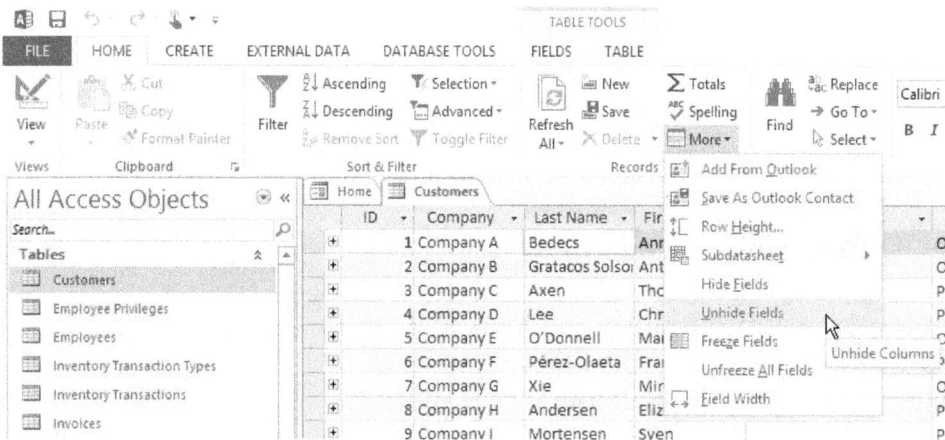

Step 3: In the Unhide Columns dialog box, check the boxes next to the fields that you would like to show again.

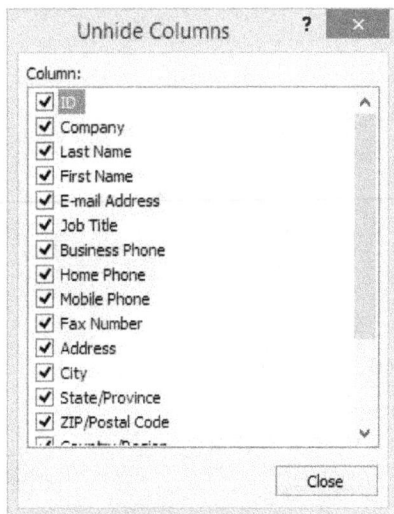

Step 4: Select Close.

Freezing and Unfreezing Fields

Rather than hiding a field in a lengthy datasheet, freezing allows you to always have a column or row showing

To freeze fields, use the following procedure.

Step 1: Open the table that you want to modify in Datasheet view.

Step 2: Move the fields that you want to freeze so that they are next to each other. You can move fields by dragging them to the new location.

Step 3: Select the fields that you want to freeze. You can hold down the SHIFT key while you select more than one field.

Step 4: Select More from the Home tab on the Ribbon.

Step 5: Select Freeze Fields.

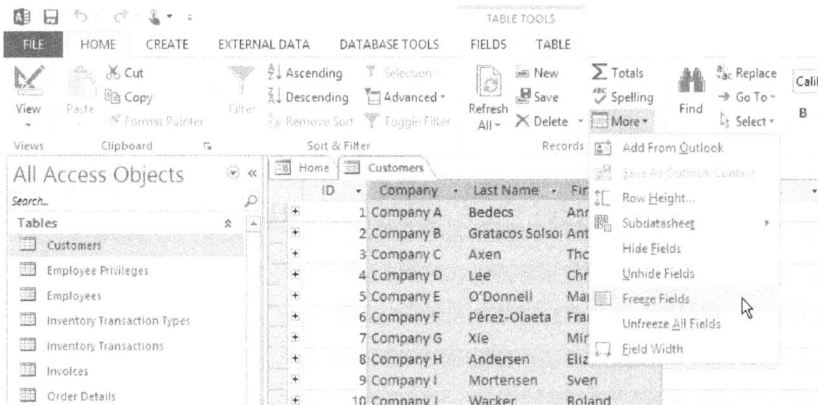

Scroll to the right to see the selected fields remain in place while the rest of the datasheet scrolls.

Step 6: If you save your changes when you close the datasheet, the fields will remain frozen until you unfreeze them.

To unfreeze fields, use the following procedure.

Step 1: Right-click on any of the column headers.

Step 2: Select Unfreeze All Fields from the context menu.

Adding from Outlook

In a database that includes contacts, you can add data from your Outlook contacts.

To add contacts from Outlook, use the following procedure.

Step 1: Open the table where you want to import the contacts.

Step 2: Select More from the Home tab on the Ribbon.

Step 3: Select Add From Outlook.

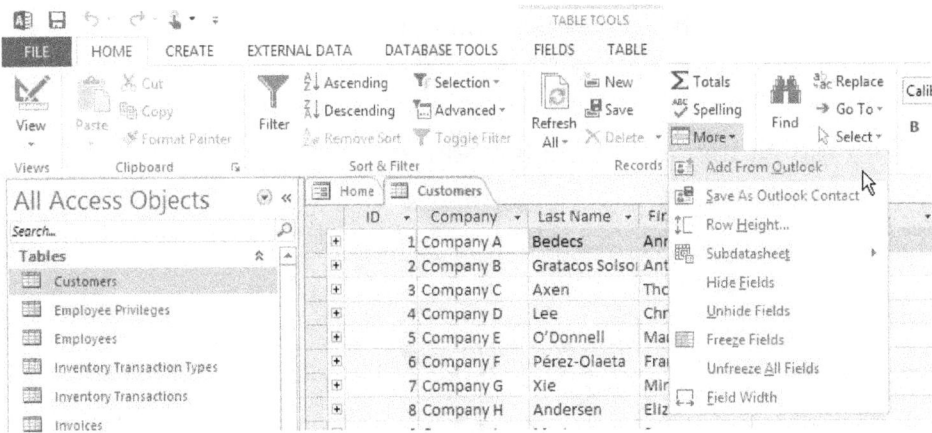

Step 4: In the Choose Profile dialog box, select the Outlook Profile Name that you want to use from the drop-down list. Select OK.

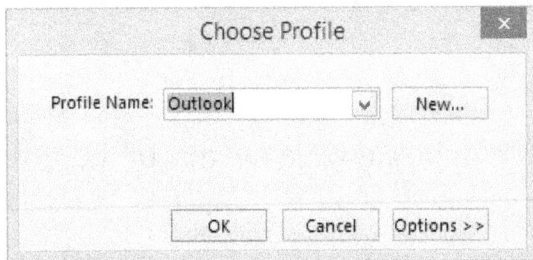

Step 5: In the Select Names to Add dialog box, highlight the contacts that you want to import. You can hold down the Shift key to select multiple contacts contiguously, or hold down the Ctrl key to select noncontiguous names. Select Add. Select OK.

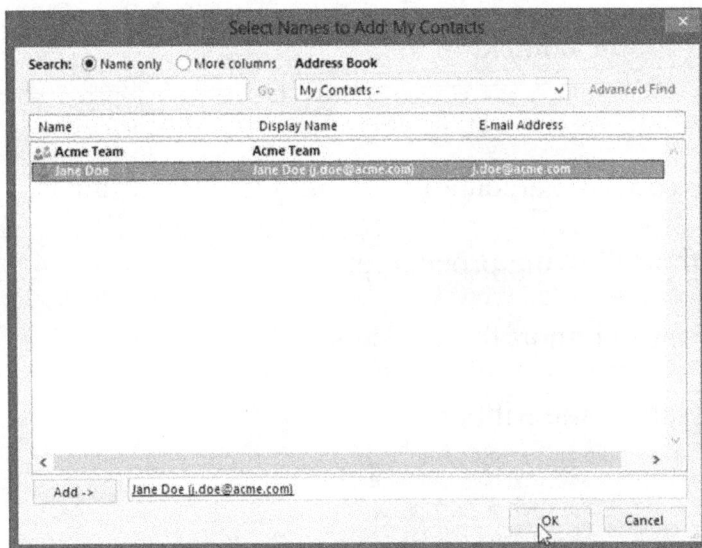

Step 6: Access displays a confirmation window when the action is complete. Select OK. Then you can view your contacts in your table.

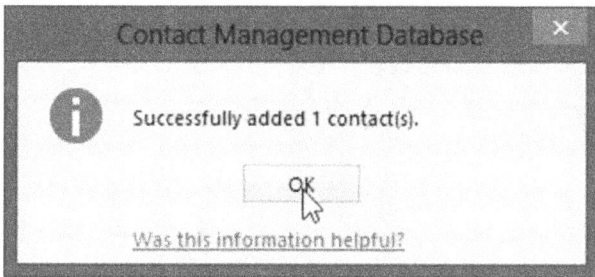

Saving a Record as an Outlook Contact

You can also export your contacts to Outlook.

To save a record as an Outlook Contact, use the following procedure.

Step 1: Select the contact that you want to save from your contacts table.

Step 2: Select More from the Home tab on the Ribbon.

Step 3: Select Save As Outlook Contact.

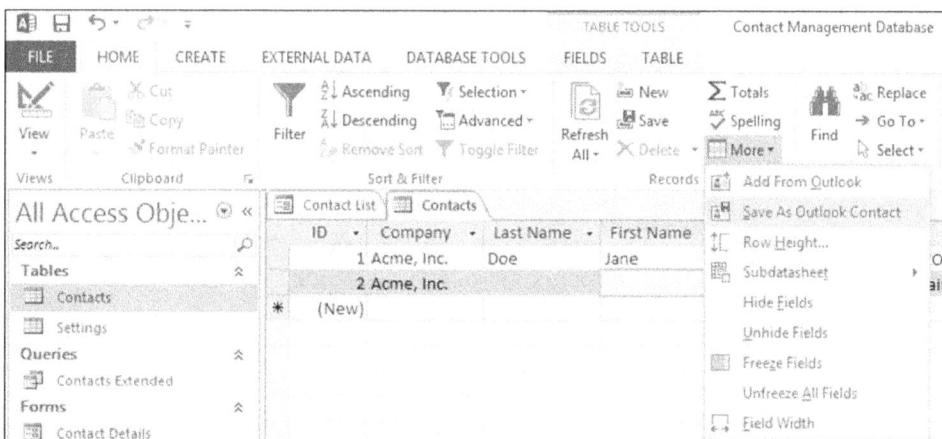

Step 4: Outlook opens a new contact. You can add information or make changes as needed. Select Save and Close.

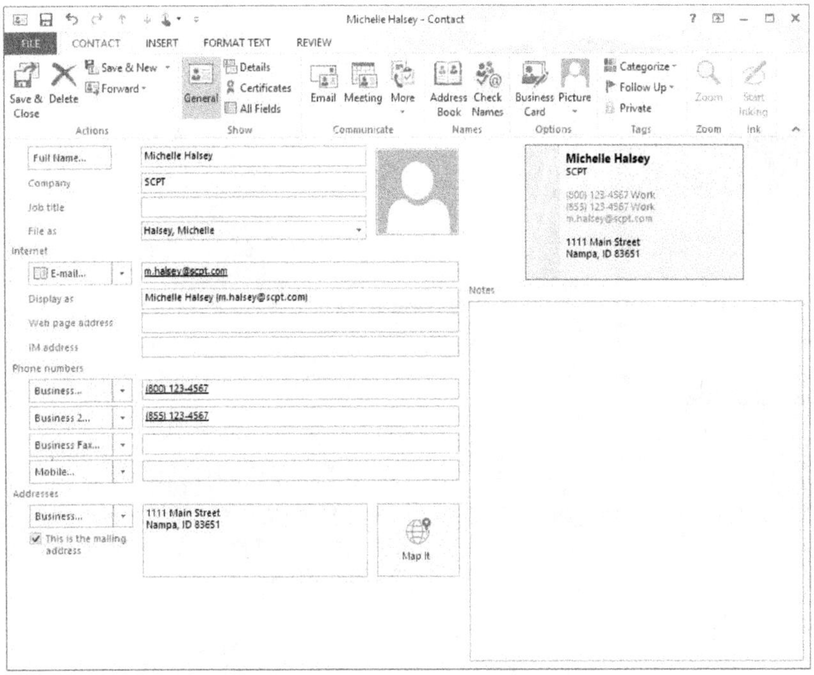
Michelle Halsey - Contact
FILE
CONTACT
INSERT
FORMAT TEXT
REVIEW
Save & Close
Delete
Save & New
Forward
Actions
General
Details
Certificates
All Fields
Show
Email
Meeting
More
Communicate
Address Book
Check Names
Names
Business Card
Picture
Options
Categorize
Follow Up
Private
Tags
Zoom
Start Inking
Ink
Full Name...
Michelle Halsey
Company
SCPT
Job title
File as
Halsey, Michelle
Internet
E-mail...
m.halsey@scpt.com
Display as
Michelle Halsey (m.halsey@scpt.com)
Web page address
IM address
Phone numbers
Business...
(800) 123-4567
Business 2...
(855) 123-4567
Business Fax...
Mobile...
Addresses
Business...
1111 Main Street
Nampa, ID 83651
This is the mailing address
Map It
Notes
Michelle Halsey
SCPT
(800) 123-4567 Work
(855) 123-4567 Work
m.halsey@scpt.com
1111 Main Street
Nampa, ID 83651

Chapter 10: Using Advanced Field Tasks

We will learn some additional tools for working with your data in this chapter. First, we will learn how to show a totals row in your datasheet. You will learn how to require fields and how to require that fields contain unique information. You will also learn about indexing fields. Finally, we will look at adding fields that are based on lookup tables and relationships.

Showing Totals

A Totals row can help you quickly identify the totals for a column on a datasheet.

To add a totals row, use the following procedure.

Step 1: From the Navigation Pane, double-click the object (table, query or split form) that you want to modify.

Step 2: Select the Home tab from the Ribbon.

Step 3: Select Totals.

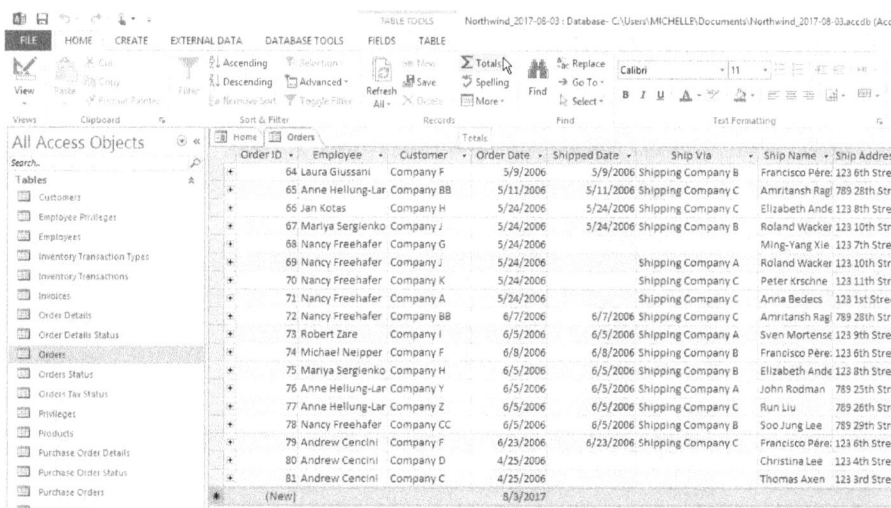

Step 4: Access adds a Total row at the end of the selected datasheet.

Step 5: In the first column, we can show the word "Total" or a count of the records in the datasheet. Click on that cell and select an option from the drop-down list.

Step 6: For each cell in the Total row where you want a total to appear, click in the cell to select the total type. The data type for the selected column must be number, currency, or decimal to select Sum. For nonnumeric columns, you can select Count.

In this example, we have added a count total to the Product column, an average total type to the Quantity column, and a sum to the Unit Price column.

Working with Required Fields and Unique Fields

We will look at the Required and Unique options in this lesson.

To make a field required, use the following procedure.

Step 1: Open the table you want to modify.

Step 2: Select the field that you want to require.

Step 3: Select the Fields tab from the Table Tools Ribbon.

Step 4: Check the Required box in the Field Validation menu.

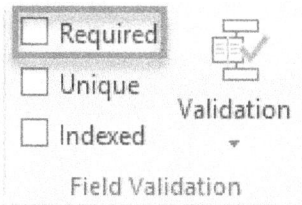

If your table has existing data that violates the Required rule, Access displays a warning message.

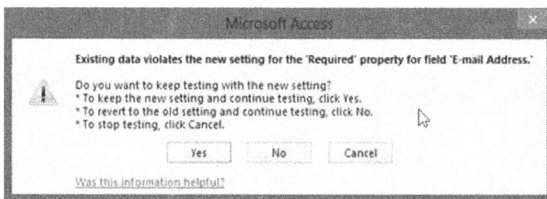

To make a field unique, use the following procedure.

Step 1: Open the table you want to modify.

Step 2: Select the field that you want to modify.

Step 3: Select the Fields tab from the Table Tools Ribbon.

Step 4: Check the Unique box in the Field Validation menu.

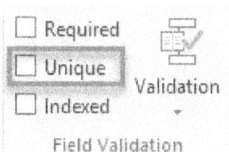

If your table has existing data that violates the Unique rule, Access displays a warning message.

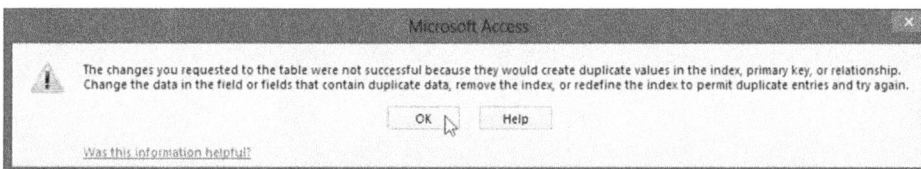

Working with Indexing

Indexes help Access find information in your data quickly.

To create a single-field index, use the following procedure.

Step 1: Open the table you want to modify.

Step 2: Select the field that you want to require.

Step 3: Select the Fields tab from the Table Tools Ribbon.

Step 4: Check the Indexed box.

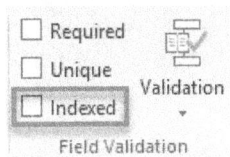

Adding Lookup and Relationship Fields

When you add fields that are based on a relationship, Access opens the Lookup & Relationship Wizard.

Creating a field in a table using the Lookup Wizard. This example shows using data entered the Lookup Wizard, use the following procedure.

Step 1: Select the arrow on the Click to Add field header.

Step 2: Select Lookup & Relationship to open the Lookup Wizard.

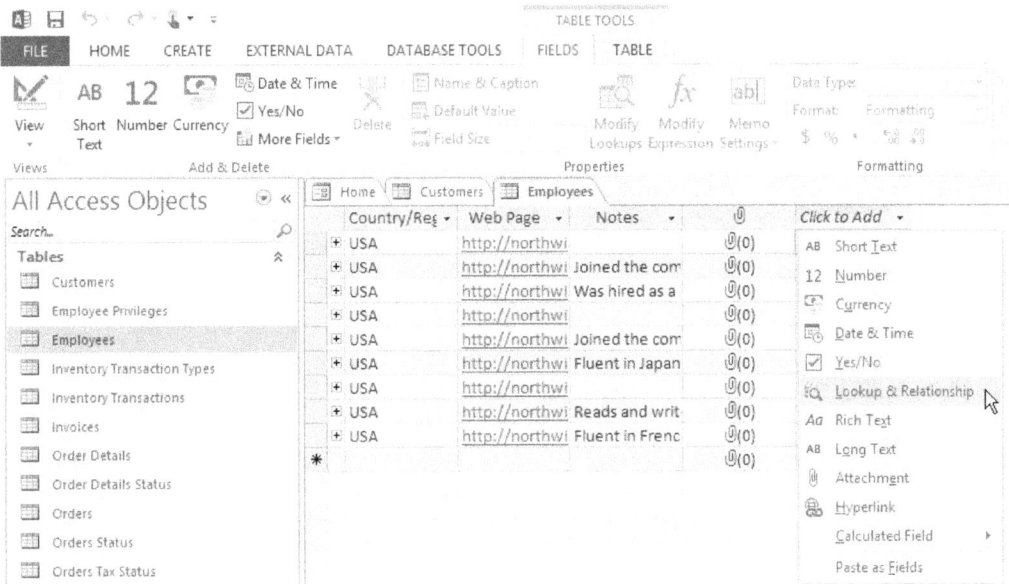

Access displays the Lookup Wizard.

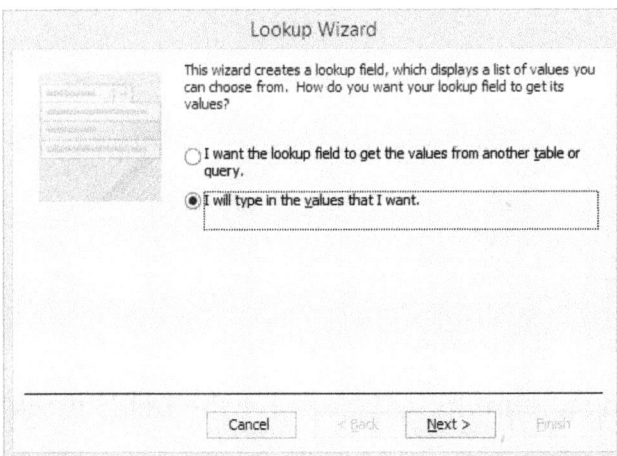

Step 3: For this first example, select I will type in the values that I want. Select Next.

Step 4: Begin entering the values. You can include multiple columns, such as to include first and last name. When you have finished entering values, select Next.

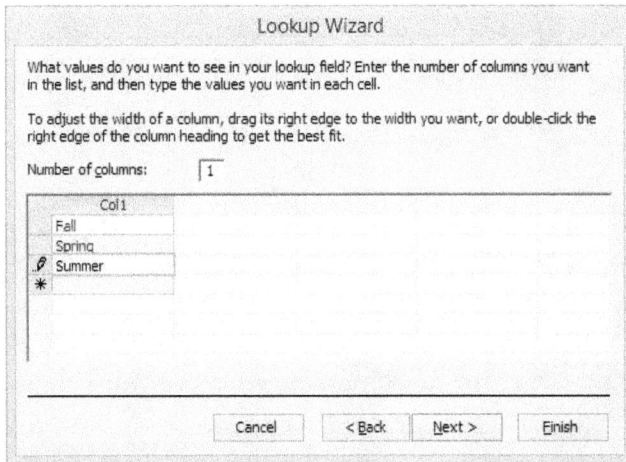

Step 5: Enter a label (field name or column heading) for your lookup column.

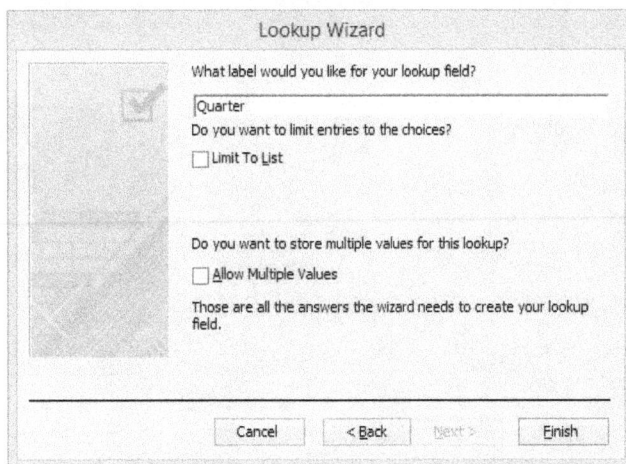

Step 6: To create a list of checkboxes, and store multiple values for the field, check the Allow Multiple Values box.

Step 7: Select Finish.

Now try selecting an option from the field you created to see the choices you entered.

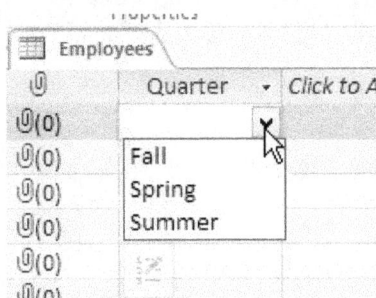

To create a field in a table using the Lookup Wizard. This example shows using data from another table, use the following procedure.

Step 1: Select Lookup & Relationship to open the Lookup Wizard.

Step 2: From the Lookup Wizard, select I want the lookup column to look up the values in a table or query. Select Next.

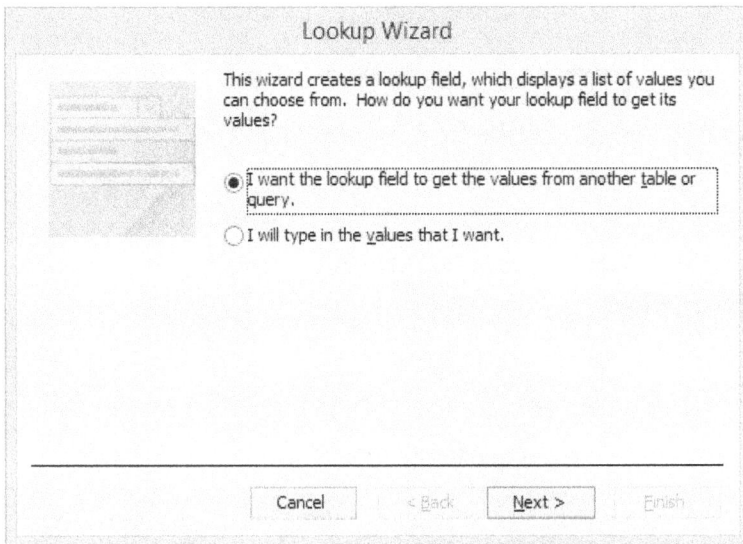

Access displays a list of tables in the current database.

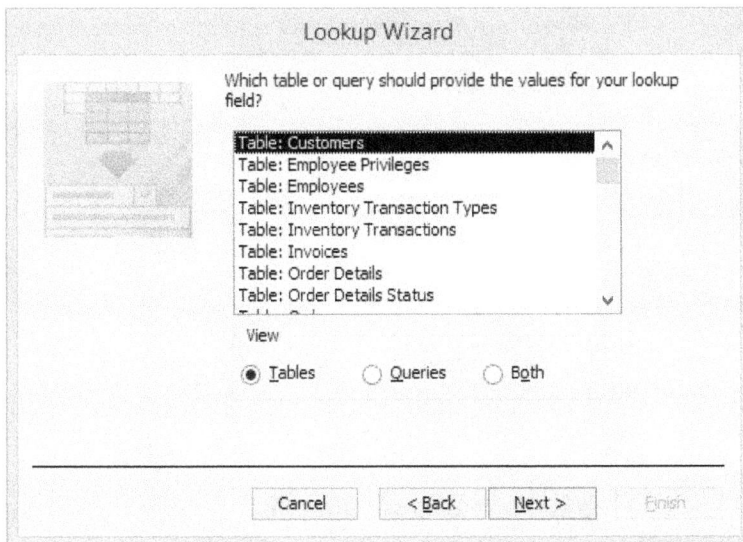

Step 3: You can view the list of possibilities by Tables, Queries, or Both. Select the table you want to use for the lookup column. Select Next.

Step 4: You can include one or more fields from the selected table in your lookup column. Each field becomes a column in the drop-down list of your new table. Highlight one or more fields and select the right arrow, or select the double right arrow to add all the fields from the table. Or you can simple double-click the field you want to select. The left arrows remove the field from the list if necessary. Select Next when you have finished.

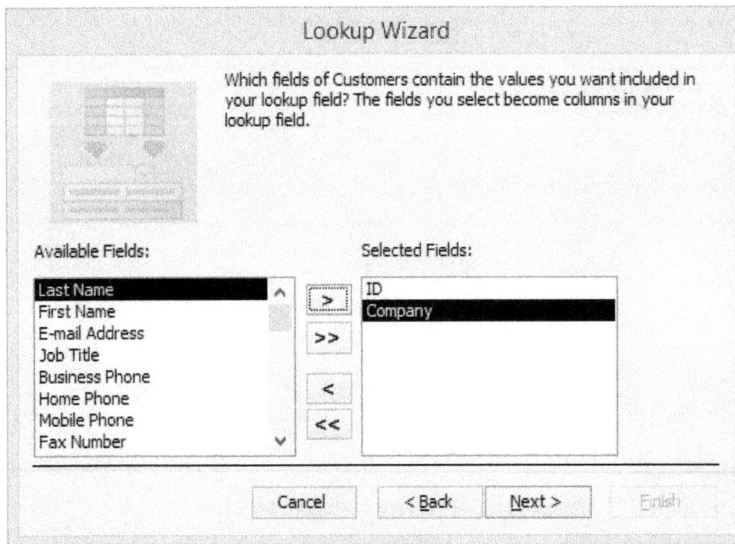

Step 5: The next screen in the wizard allows you to determine the order the values appear in the drop-down list.

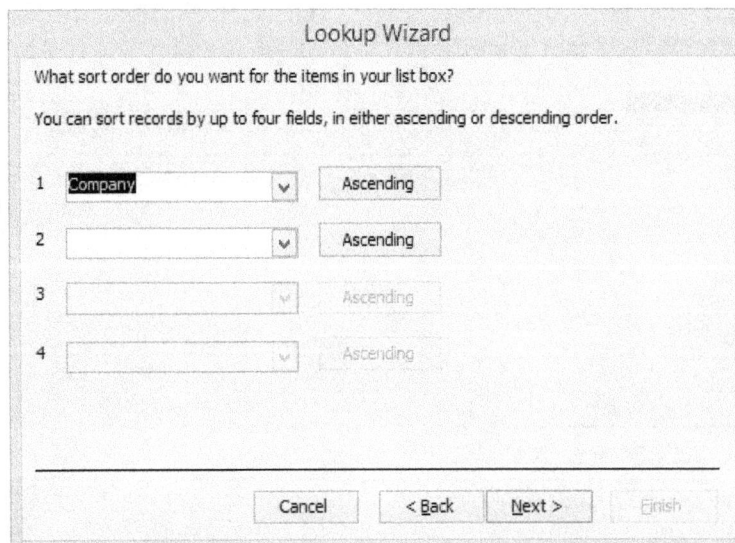

Step 6: Select the field for each of up to four choices. Select Ascending or Descending by clicking the button. Select Next.

Step 7: The next screen in the Lookup Wizard allows you to control the width of the column(s) in your drop-down list. You can drag the column(s) wider or narrower to suit the values. When you have finished, select Next.

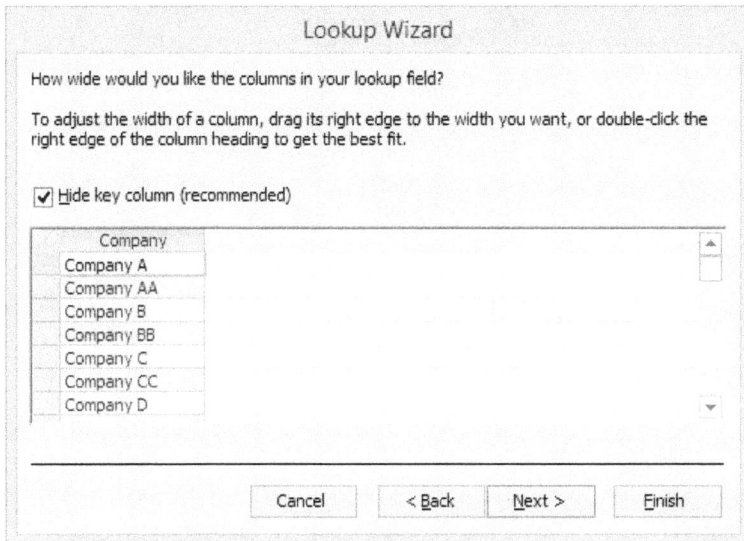

Lookup Wizard

How wide would you like the columns in your lookup field?

To adjust the width of a column, drag its right edge to the width you want, or double-click the right edge of the column heading to get the best fit.

☑ Hide key column (recommended)

Company
Company A
Company AA
Company B
Company BB
Company C
Company CC
Company D

Cancel   < Back   Next >   Finish

Step 8: The next screen in the Lookup Wizard allows you to name the field or column heading. Enter the name and select Finish.

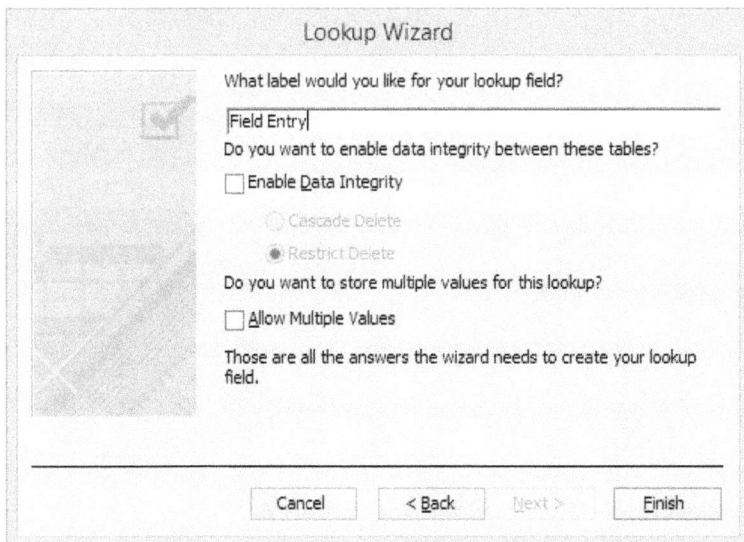

Lookup Wizard

What label would you like for your lookup field?

Field Entry

Do you want to enable data integrity between these tables?

☐ Enable Data Integrity

○ Cascade Delete
● Restrict Delete

Do you want to store multiple values for this lookup?

☐ Allow Multiple Values

Those are all the answers the wizard needs to create your lookup field.

Cancel   < Back   Next >   Finish

Chapter 11: Working in Table Design View

This chapter will introduce you to the Design view for tables. You will learn how to work with the field properties and the primary key. You will also learn how to use the Properties Sheet.

Opening Design View

Design view is a view where you can set many details and properties for your table at once.

To create a table using Design View, use the following procedure.

Step 1: Select the Create tab from the Ribbon.

Step 2: Select Table Design.

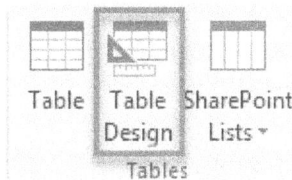

Step 3: For each field, you want to include in the table, enter the following information:

- Field Name
- Data Type
- Description (optional)

The following example shows the Design view of the Inventory Transactions table from the Northwind sample database. This example includes several different data types. A description is added as a prompt for table users. The description will appear in the status bar when the table is being used in Datasheet view.

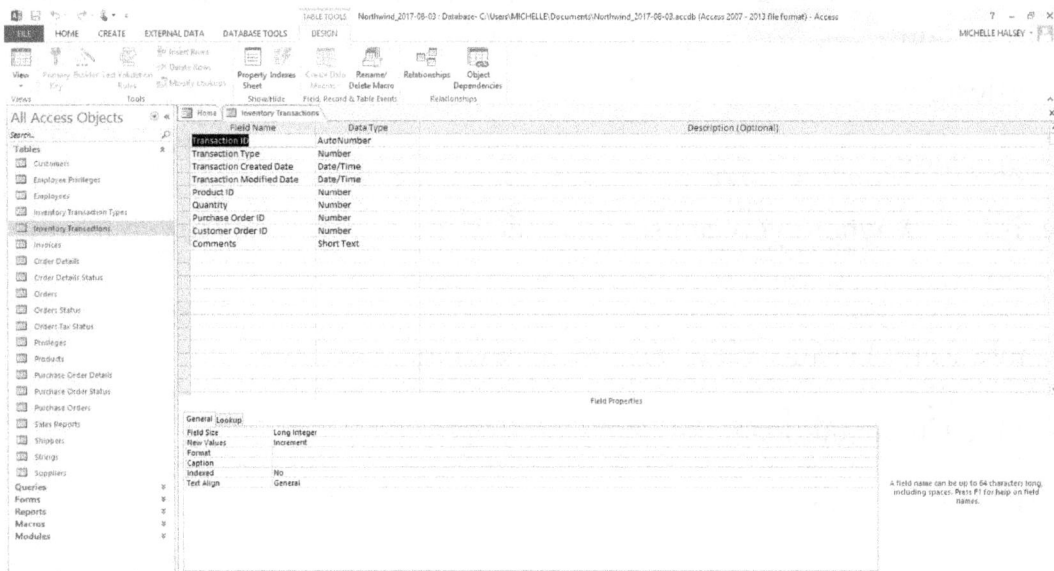

Setting the Primary Key

The primary key is the unique identifier for each table.

To set the Primary Key, use the following procedure.

Step 1: To open a table in Design View, select View from the Home tab on the Ribbon.

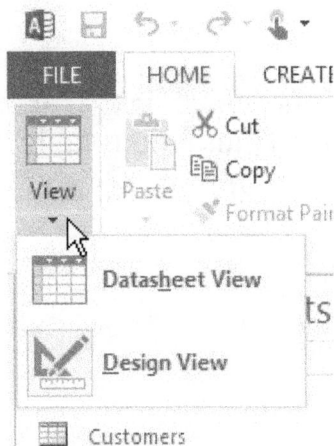

Step 2: Select the Field Name that you want to set as the Primary Key.

Step 3: Select Primary Key from the Table Tools Design tab on the Ribbon.

86

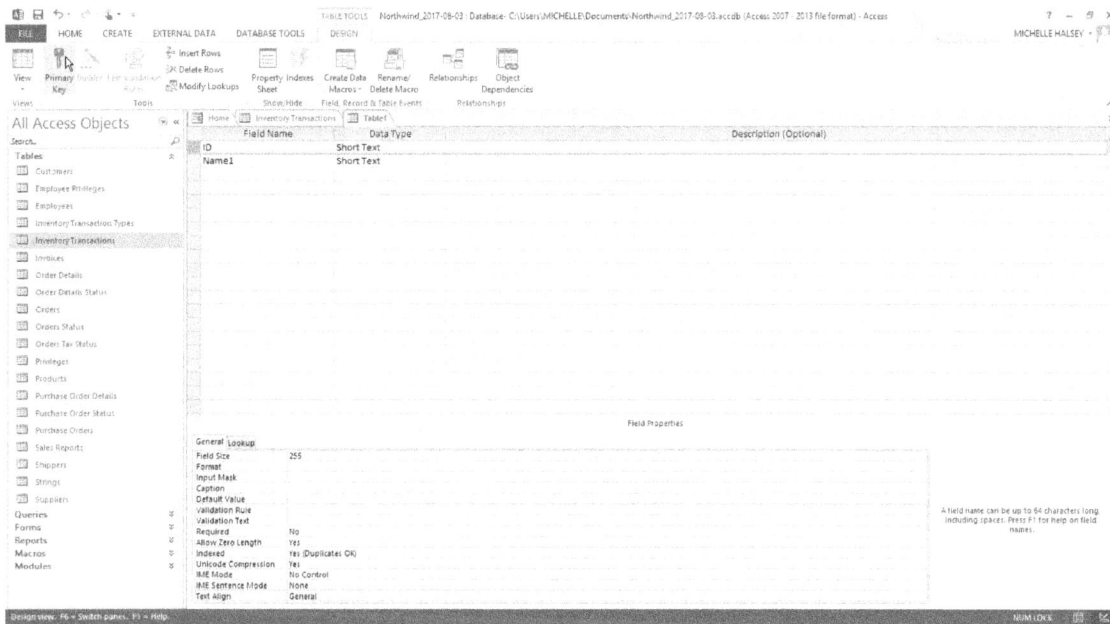

Step 4: Access adds a key icon next to the field that you have set as the Primary Key.

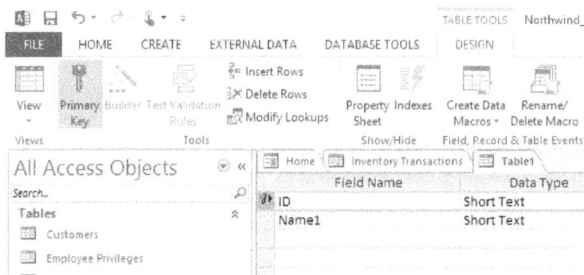

Working with Field Properties

The Field Properties area in Design view allows you to change all the properties for a field at one time. Each data type has different rules on the formats that can be used, the maximum size of the field, how the field can be used in expressions, and whether the field can be indexed. The following data types are supported:

- Text
- Number
- Currency
- Yes/No
- Date/Time
- Rich Text

- General
- Currency
- Euro
- Fixed
- Standard
- Percentage

- Medium Time
- Time 24hour
- Check Box
- Yes/No
- True/False
- On/Off

- Calculated Field
- Attachment
- Hyperlink
- Memo
- Lookup

- Scientific
- Short Date
- Medium Date
- Long Date
- Time am/pm

- Address
- Phone
- Priority
- Status
- Tags

The field's data type determines which other properties you can set. Refer to the Appendix for a brief reference of which field properties are included with which data types.

To view the field properties, use the following procedure.

Step 1: Open the table that you want to view in Design View. If it is already open in Datasheet view, right-click on the tab for that table and select Design View from the context menu.

Step 2: The list of fields appears at the top of the screen. The Field Properties are shown at the bottom. When you select a different field, the field properties for the selected field are displayed. Note that help for completing the current field property is shown to the right.

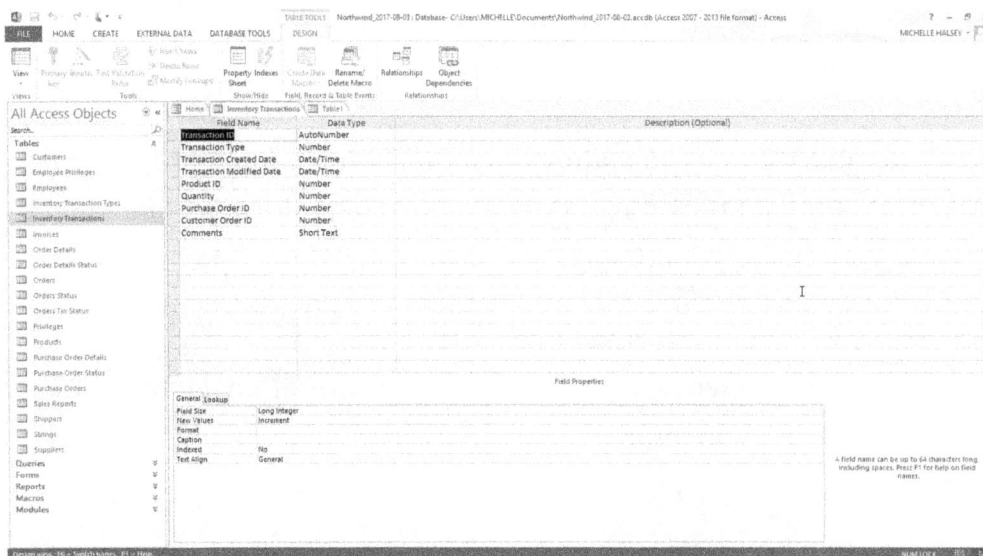

Let's practice changing the field size in the Customers table.

Step 1: Select the ZIP/Postal code field.

Step 2: In the Field Size area under Field Properties, change 15 to 9.

Practice viewing the field properties for different types of data types. Use the Orders table in the Northwind database.

Review the Field Properties section of the Design View for several different types of data fields.

From the above example, these are the Field Properties from an Auto Number field.

General	Lookup
Field Size	Long Integer
New Values	Increment
Format	
Caption	
Indexed	Yes (No Duplicates)
Text Align	General

These are the field properties for a Number data type.

General	Lookup
Field Size	Long Integer
Format	
Decimal Places	Auto
Input Mask	
Caption	Employee
Default Value	
Validation Rule	
Validation Text	
Required	No
Indexed	Yes (Duplicates OK)
Text Align	General

These are the field properties from a Date/Time data type.

General	Lookup
Format	Short Date
Input Mask	
Caption	
Default Value	
Validation Rule	
Validation Text	Value must be greater than 1/1/1900.
Required	No
Indexed	No
IME Mode	Off
IME Sentence Mode	None
Text Align	General
Show Date Picker	For dates

Review additional field properties as time permits.

Using the Properties Sheet

This lesson introduces the Properties Sheet, which is available in Design view for tables and other objects.

To open the Properties Sheet, use the following procedure.

Step 1: Select the Design tab from the Table Tools Ribbon.

Step 2: Select Property Sheet in the Show/Hide menu.

Step 3: Discuss the properties available in the Properties Sheet when working with a table.

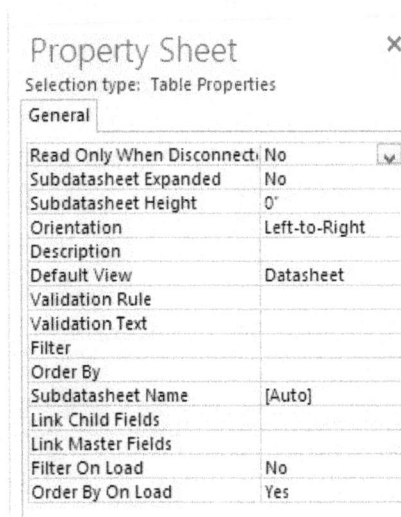

Click on one of the fields to change its value or to add a value if one does not exist.

Chapter 12: Working with the Expression Builder

This chapter focuses on the Expression Builder, and ways that you can use it to improve your database. We will start by looking at setting the default value for a field, both in design view and in datasheet view. Then we will take a closer look at the Expression builder to create formulas for all different types of uses in Access. Finally, we will create calculated value fields.

Setting the Field Default Value

You can create a default value for your fields.

To set the default value in Design View, use the following procedure.

Step 1: Open the table in Design View.

Step 2: Find the field where you want to include a default value. In this example, we will use the Customers table, as if all our customers usually came from the same state.

Step 3: In the Field Properties, enter the Default Value.

Field Name	Data Type
ID	AutoNumber
Company	Short Text
Last Name	Short Text
First Name	Short Text
E-mail Address	Short Text
Job Title	Short Text
Business Phone	Short Text
Home Phone	Short Text
Mobile Phone	Short Text
Fax Number	Short Text
Address	Long Text
City	Short Text
State/Province	Short Text
ZIP/Postal Code	Short Text
Country/Region	Short Text
Web Page	Hyperlink
Notes	Long Text
Attachments	Attachment

General Lookup

Field Size	50
Format	
Input Mask	
Caption	
Default Value	TX
Validation Rule	
Validation Text	
Required	No
Allow Zero Length	No
Indexed	Yes (Duplicates OK)
Unicode Compression	Yes
IME Mode	No Control
IME Sentence Mode	Phrase Predict
Text Align	General

Step 4: Notice that when you click elsewhere or press TAB to move to the next property, Access adds quotation marks around the string (if you did not).

General	Lookup	
Field Size	50
Format	
Input Mask	
Caption	
Default Value	|TX"
Validation Rule	
Validation Text	
Required	No
Allow Zero Length	No
Indexed	Yes (Duplicates OK)
Unicode Compression	Yes
IME Mode	No Control
IME Sentence Mode	Phrase Predict
Text Align	General

To set the default value in Datasheet View, use the following procedure.

Step 1: Open the table in Datasheet view.

Step 2: Select the field where you want to include a default value.

Step 3: Select the Fields tab from the Table Tools Ribbon.

Step 4: Select Default Value in the Properties menu.

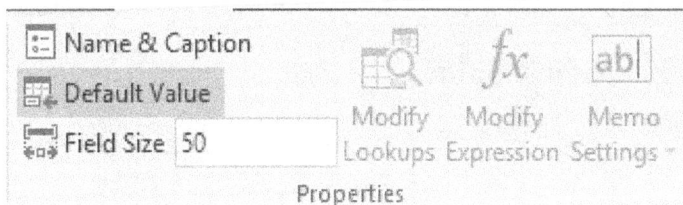

☷ Name & Caption			*fx*	ab|
⊞ Default Value		Modify	Modify	Memo
⊟ Field Size 50		Lookups	Expression	Settings
	Properties			

Step 5: Access opens the Expression Builder. In this example, we will just use the exact same text from the previous example. In other words, if you want an exact string, just enter it with quotation marks around it. Select OK.

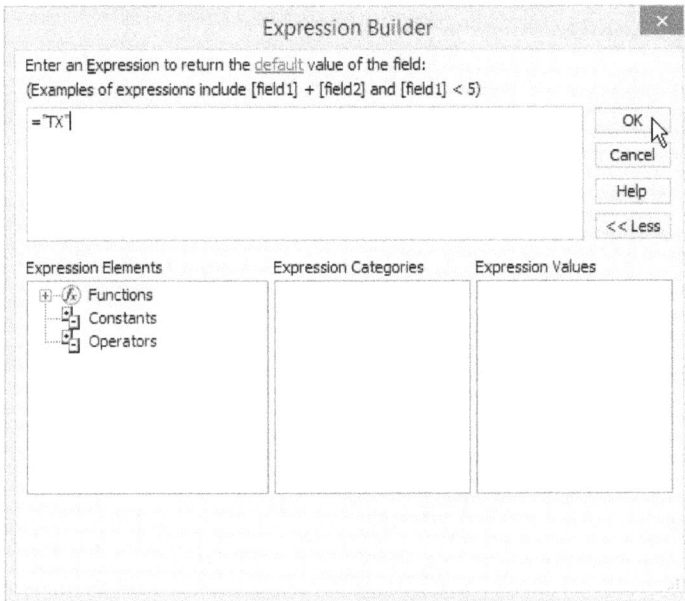

Expression Builder

Enter an Expression to return the default value of the field:
(Examples of expressions include [field1] + [field2] and [field1] < 5)

= "TX"

OK
Cancel
Help
<< Less

Expression Elements
- Functions
- Constants
- Operators

Expression Categories

Expression Values

	Home Phon… ▾	Mobile Phor ▾	Fax Number ▾	Address ▾	City ▾	State/Provir ▾	ZIP/Postal C ▾
+			(123)555-0101	123 1st Street	Seattle	WA	99999
+			(123)555-0101	123 2nd Street	Boston	MA	99999
+			(123)555-0101	123 3rd Street	Los Angelas	CA	99999
+			(123)555-0101	123 4th Street	New York	NY	99999
+			(123)555-0101	123 5th Street	Minneapolis	MN	99999
+			(123)555-0101	123 6th Street	Milwaukee	WI	99999
+			(123)555-0101	123 7th Street	Boise	ID	99999
+			(123)555-0101	123 8th Street	Portland	OR	99999
+			(123)555-0101	123 9th Street	Salt Lake City	UT	99999
+			(123)555-0101	123 10th Street	Chicago	IL	99999
+			(123)555-0101	123 11th Street	Miami	FL	99999

Step 6: Notice the default value entered for a new row. Point out that even though the default value is automatically there for a new record, it can be changed.

In the next lesson, you will learn how to write formulas with the Expression Builder. Note that when setting the default value in Design View, you can simply enter any formula as you might create in the Expression Builder. You can also select the three dots to the right of the Default Value field to open the Expression Builder.

Using the Expression Builder

The Expression Builder helps you to create an expression – the Access version of formulas.

To open the Expression Builder, click the three dots button in the input field.

Here is the layout of the Expression Builder.

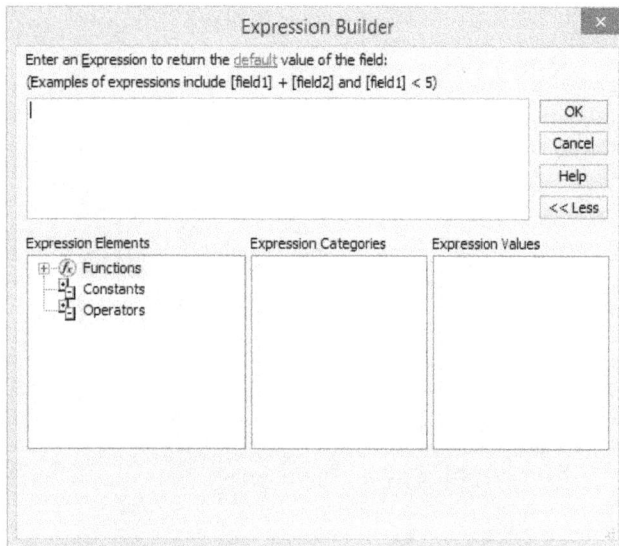

The top area of the Expression builder is for instructions and a help link. Access displays information about the context in which you are entering an expression.

The Expression box is where you enter your expression. You can use the elements listed below. If they are not visible, select More.

The Expression Elements shows the elements you can use in your expression. Click an item to view its categories.

The Expression Categories List shows categories. Click an item to show its values or double-click it to add it to the expression.

The Expression Values list shows the values for the selected element and category. Double-click a value to add it to the expression.

Remember that you can type your expression manually in the Expression box or you can use the tools below to help you build your expression.

Here is more information on the tools to help you build an expression.

- The Expression Elements are the top-level elements, such as database objects, functions, constants, operators, and common expressions.
- The Expression Categories list contains specific categories of elements for the selection that you make in the Expression Elements list.

- The Expression Values list contains specific values for the element and category you selected in the left and middle lists.

To add an element to the Expression Builder, use the following procedure.

Step 1: Select an item in the Expression Elements list. For example, to insert a built-in function, expand Functions and then select Built-in Functions.

Step 2: Select an item in the Expression Categories. In this example, we will select Date/Time.

Step 3: Select an item in the Expression Values list. In this example, double-click Now in the Expression values list to add it to the Expression box.

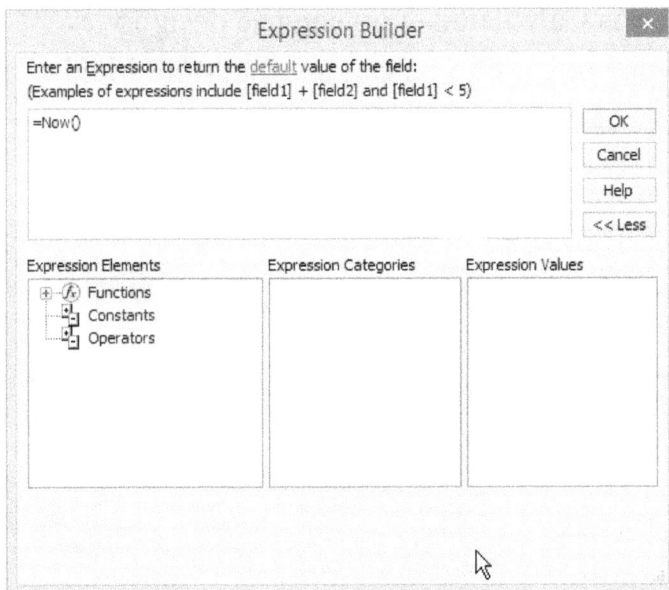

Step 4: For more complex functions, you can replace any placeholder text with the valid argument values. For example, look at before and after for the following Month example.

Before	=Month(<<date>>)
After	=Month(5)

Step 5: If the expression contains other elements, they may be separated by the placeholder <<Expr>>. Replace this placeholder with an operator before the overall expression will be valid.

Step 6: Select OK to close the Expression Builder.

Adding Calculated Fields

When you add a field with the data type "calculated", Access opens the Expression Builder.

To add a calculated field, use the following procedure. In this example, we will add a total column to the order details table to calculate the order total.

Step 1: Add a field and set the data type as Calculated. You can use design view and select the data type from the drop-down list. Or you can use the More Fields option in Datasheet view.

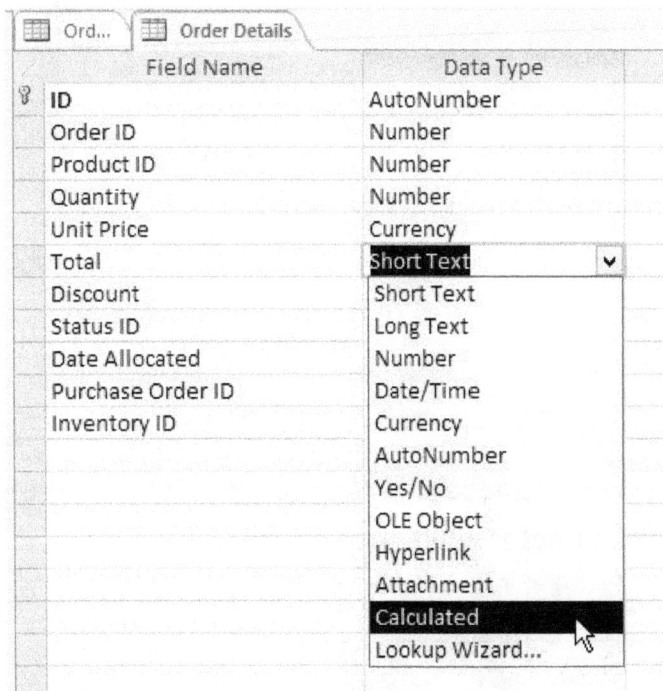

Field Name	Data Type
ID	AutoNumber
Order ID	Number
Product ID	Number
Quantity	Number
Unit Price	Currency
Total	Short Text
Discount	Short Text
Status ID	Long Text
Date Allocated	Number
Purchase Order ID	Date/Time
Inventory ID	Currency
	AutoNumber
	Yes/No
	OLE Object
	Hyperlink
	Attachment
	Calculated
	Lookup Wizard...

Step 2: We want the expression [Quantity] * [Unit Price]. You can either type this in manually, or select the field names using the tools below.

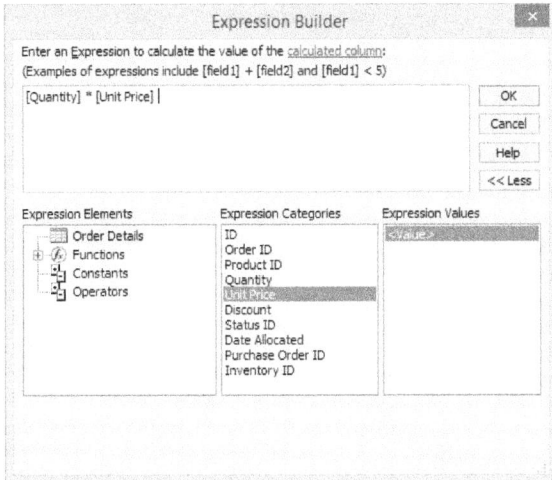

Step 3: Select OK.

Step 4: Save the table.

Let's open it in Datasheet view to view the results. There is our new column with the calculated values. We could now reformat the column, as needed.

Chapter 13: Working with External Data

This chapter helps you understand how to work with external data. We will use Excel for our examples. We will start with linking data, which keeps your data updated in Access, allowing you to use the query and reporting tools, but does not allow you to make changes to the data in Excel. Then we will look at importing data, which allows you to make changes to the data, but does not provide any updates if changes are made in Excel. Finally, we will shift to exporting data.

Linking Data

This lesson looks at linking to an Excel spreadsheet.

To link an Excel spreadsheet, use the following procedure.

Step 1: Select the External Data tab from the Ribbon.

Step 2: Select the arrow under the New Data Source Drop-down in the Import & Link menu on the External Data ribbon.

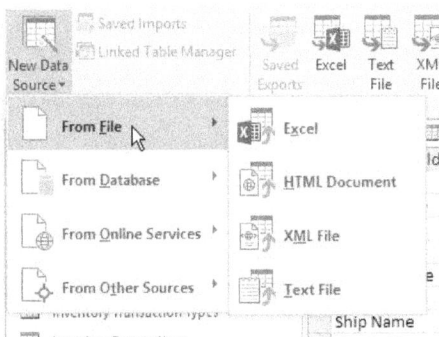

Step 3: In the Get External Data dialog box, select Browse.

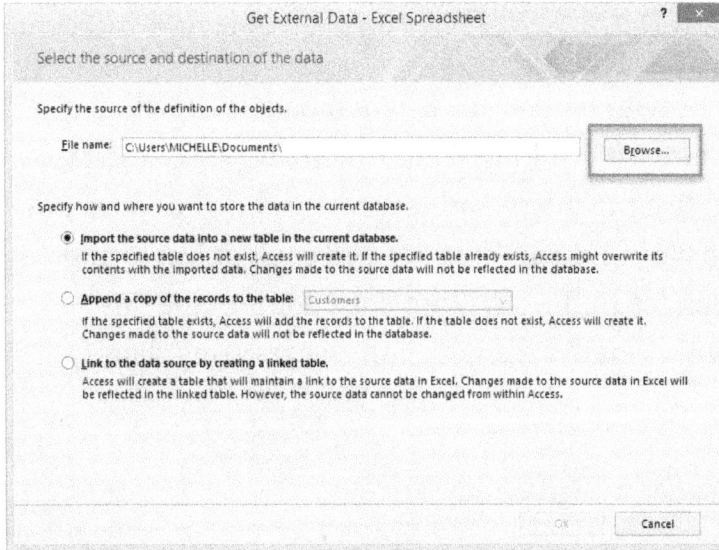

Step 4: Navigate to the location of the Excel file you want to use. Select Open.

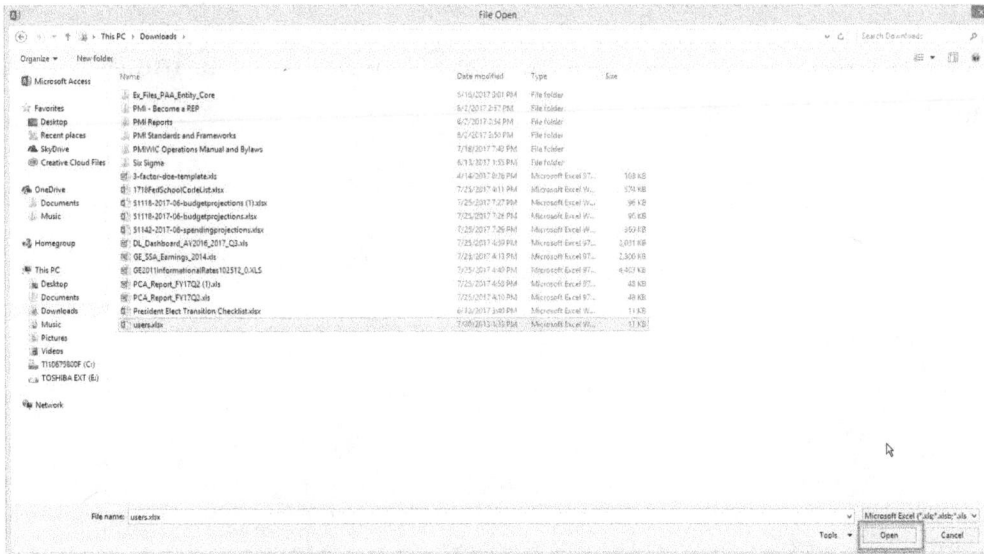

Step 5: Make sure the Link to the Data Source by Creating a Linked Table option is selected. Select OK.

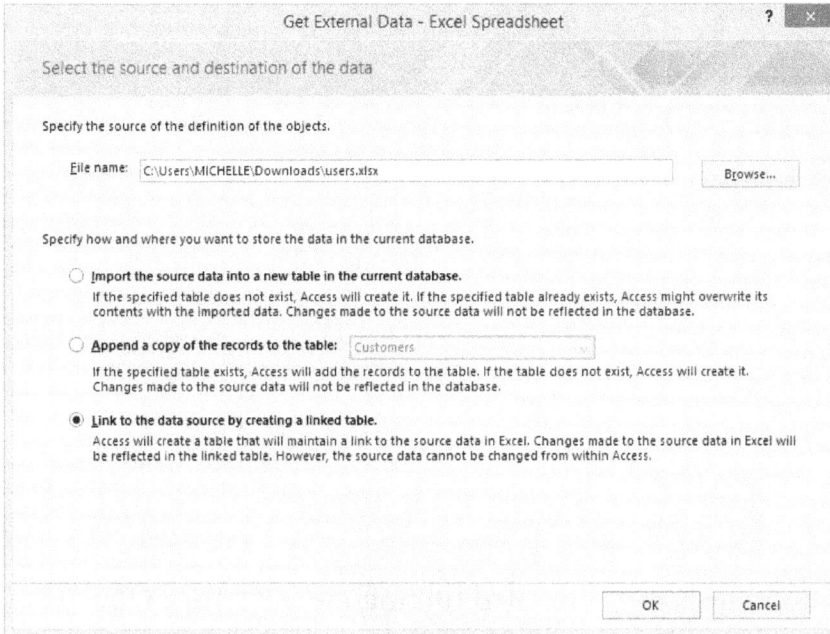

Step 6: In the Link Spreadsheet dialog box, check the First Row Contains Column Headings, if applicable. Select Next.

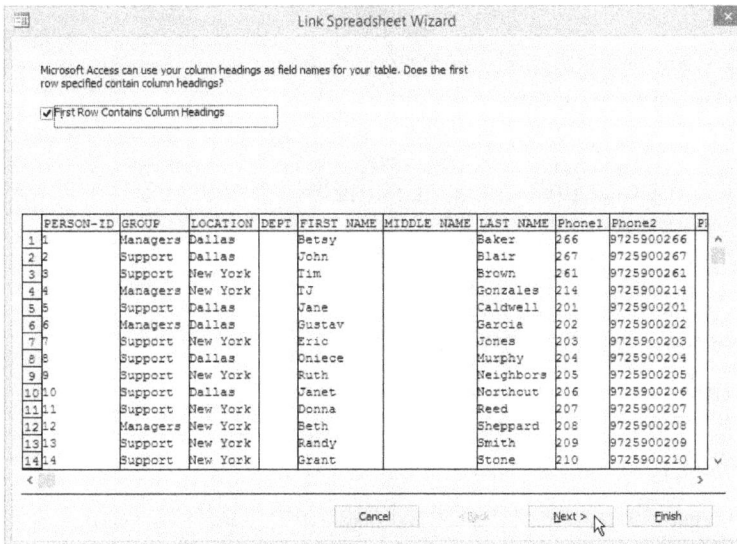

Step 7: Enter a name for the table that Access will create in the Linked Table Name field. Select Finish.

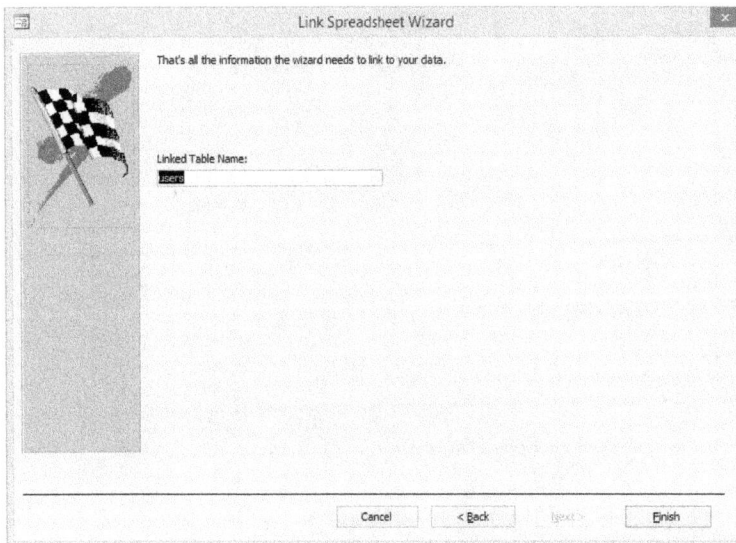

Access displays a confirmation message, "Finished linking table to file". Click Ok.

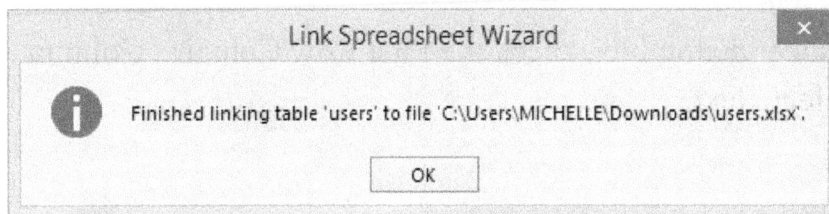

Importing Data

Importing is like linking in Access, but changes are not linked to the source.

To import an Excel spreadsheet, use the following procedure.

Step 1: Prepare the database where you want to import the table. The database must not be read-only and you should have the permissions to make changes.

Step 2: Decide whether you want to create a new table for the imported data or to add the data to an existing table. If you choose to append the data to an existing table, the Access table must have the same structure as the imported table, including headings, missing or extra fields, the primary key, and indexed fields.

Step 3: Select the External Data tab from the Ribbon.

Step 4: Select Excel.

Step 5: In the Get External Data dialog box, select Browse.

Step 6: Navigate to the location of the spreadsheet you want to import. Select Open.

Step 7: Select one of the following based on your decision in step 2:

Import the source data into a new table in the current database

Append a copy of the records to the table – If you select this option, also select the table from the drop-down list.

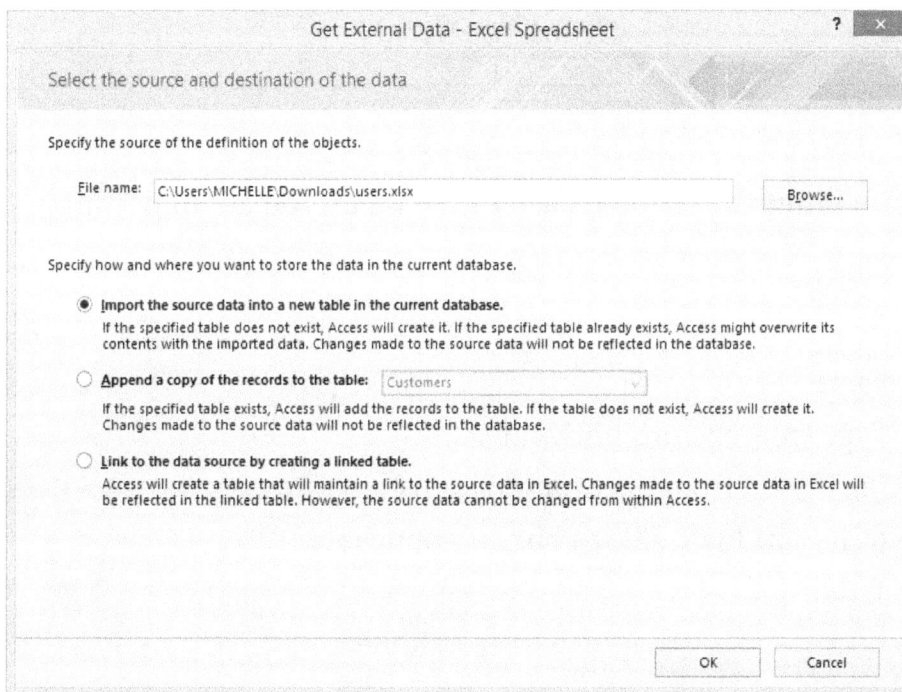

Step 8: Select OK.

Step 9: In the Link Spreadsheet dialog box, check the First Row Contains Column Headings, if applicable. Select Next.

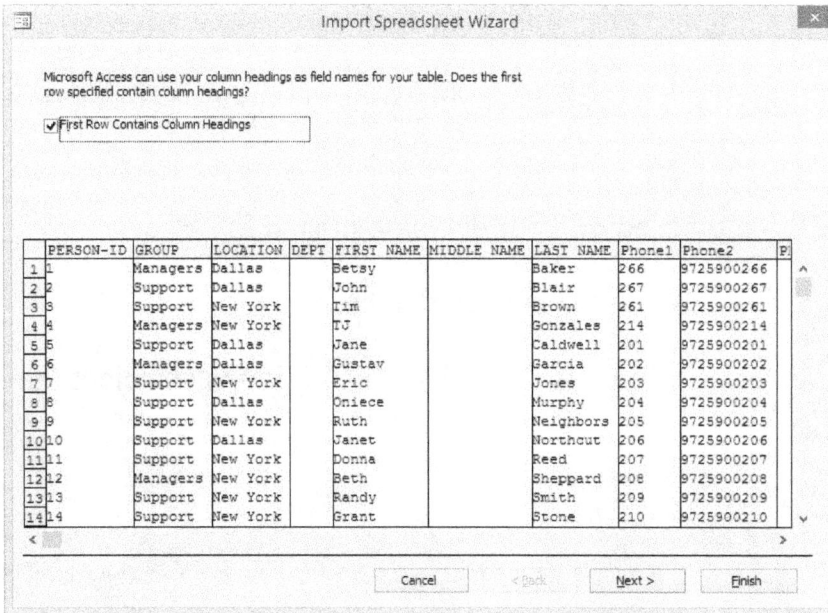

Step 10: You can specify information about each of the fields you are importing. Follow these steps:

- Select the field.
- Enter the Field Name.
- Select the Data Type from the drop-down list.
- Select the Indexing option from the drop-down list.
- Check the Do Not Import Field (Skip) box, if applicable.
- Repeat for each field.

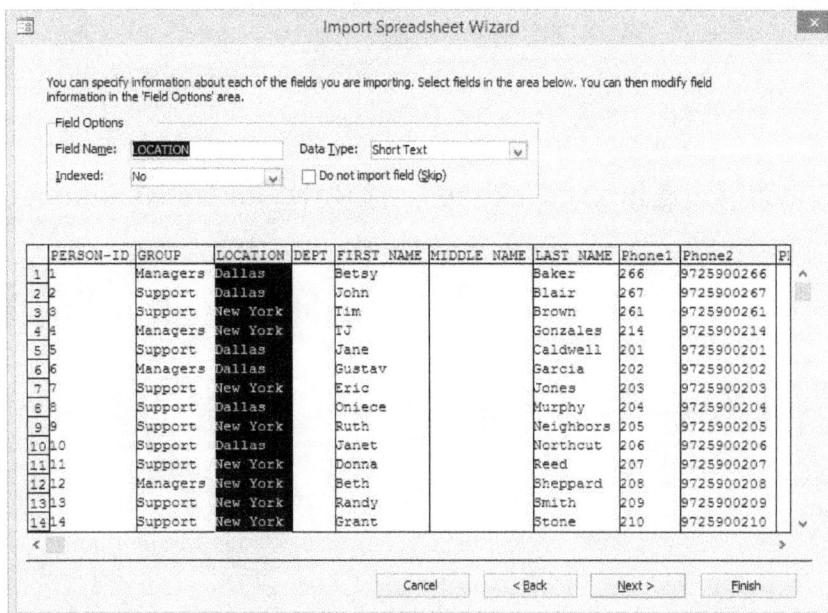

Step 11: Select a Primary Key option. If you select Choose my own primary key, select one from the drop-down list.

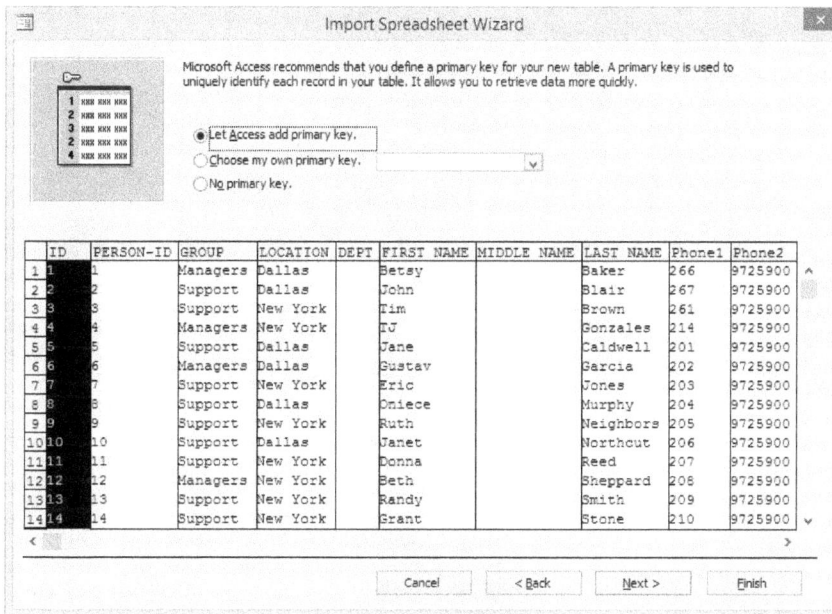

Step 12: Enter a name for the table that Access will create in the Import to Table field. Select Finish.

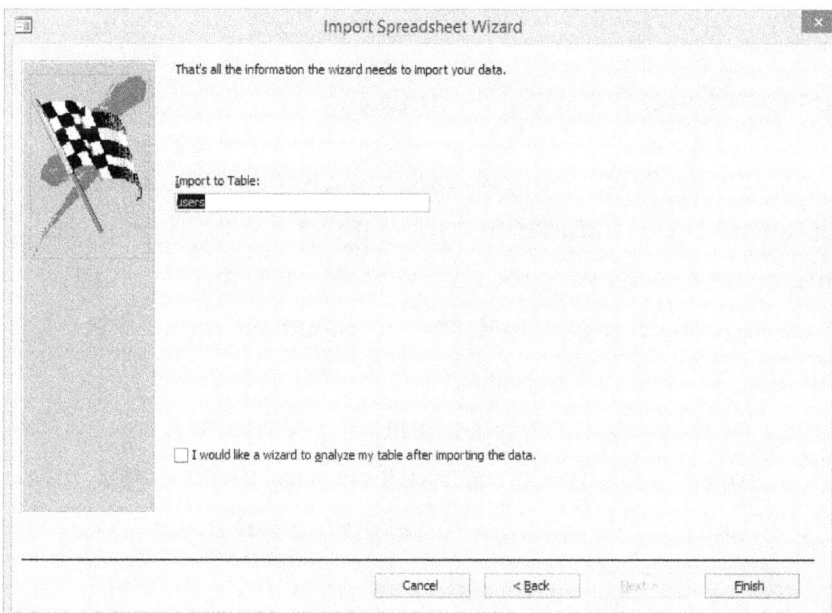

Step 13: You can save the import steps if desired by checking the Save Import Steps box. Select Close.

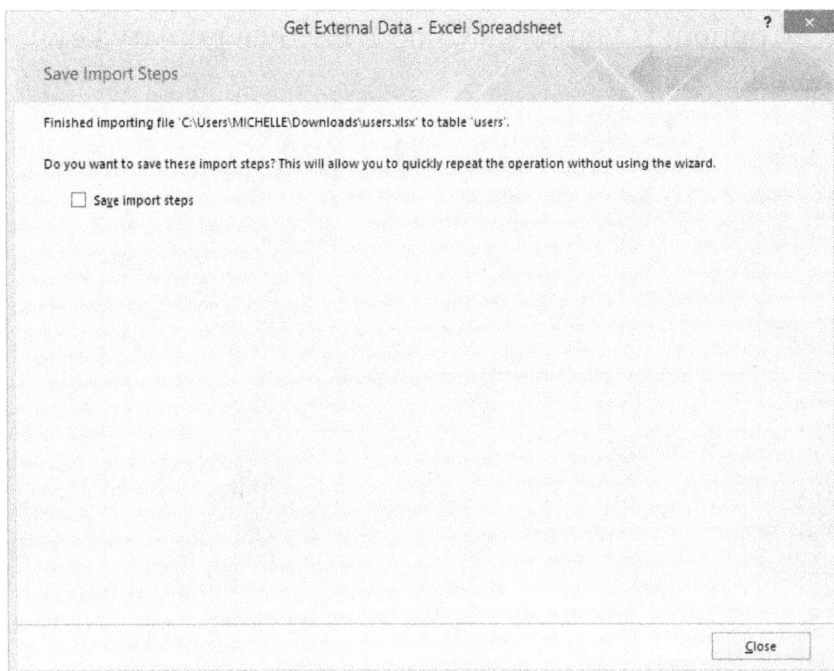

About Exporting Data

The Export function allows you to export data that can be read by Excel.

Exporting to Excel creates a copy of the selected data that can be opened in Access.

Here are some common scenarios for exporting data to Excel:

- Your department or workgroup uses both Access and Excel to work with data. You store the data in Access databases, but you use Excel to analyze the data and to distribute the results of your analysis. Your team currently exports data to Excel as and when they must, but you want to make this process more efficient.
- You are a long-time user of Access, but your manager prefers to work with data in Excel. At regular intervals, you do the work of copying the data into Excel, but you want to automate this process to save yourself time.

Here are some tips for planning your exports:

- Note that you cannot simply perform a "save as" to get your data to Excel. You must use the export feature. However, you can copy data to the clipboard from Access and paste it into an Excel spreadsheet.

- You can export tables, query data, forms, or reports. You can export selected records from a datasheet view.
- Microsoft Excel includes an import data command to retrieve data from Access. The difference between the two is that importing from Excel only allows you to import tables or queries.
- You cannot export macros or chapters to Excel. If your forms, reports, or datasheets include sub-forms, sub-reports, or sub-datasheets, those items must be exported individually.
- Each database object is exported separately. You can merge multiple worksheets in Excel once you have completed the export operations.

Exporting Data to Excel

To export data to Excel, use the following procedure.

Step 1: Open the object that you want to export.

Step 2: Select the External Data tab from the Ribbon.

Step 3: Select Excel from the Export area.

Step 4: In the Export – Excel Spreadsheet dialog box, enter the File Name. You can choose Browse to navigate to the location and select or enter the file name (depending on whether you will overwrite a file or create a new one).

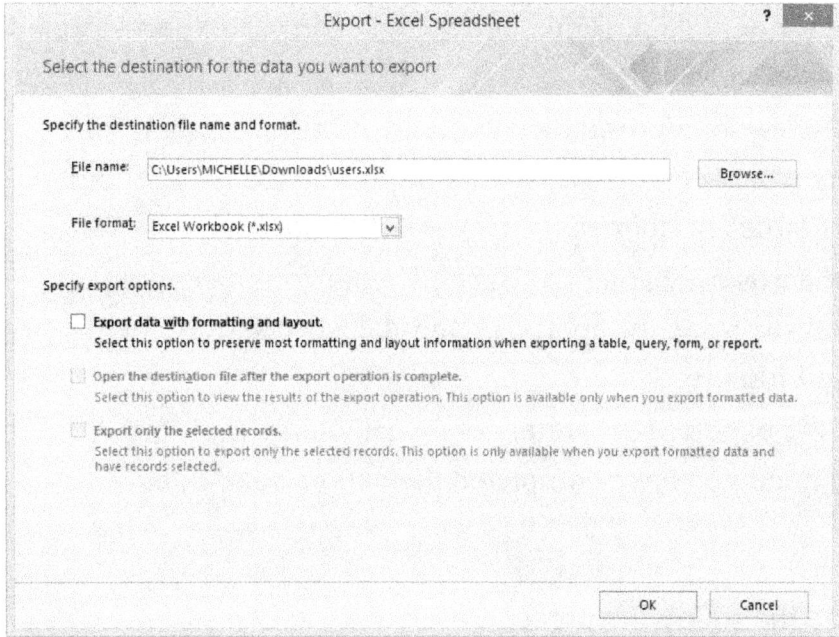

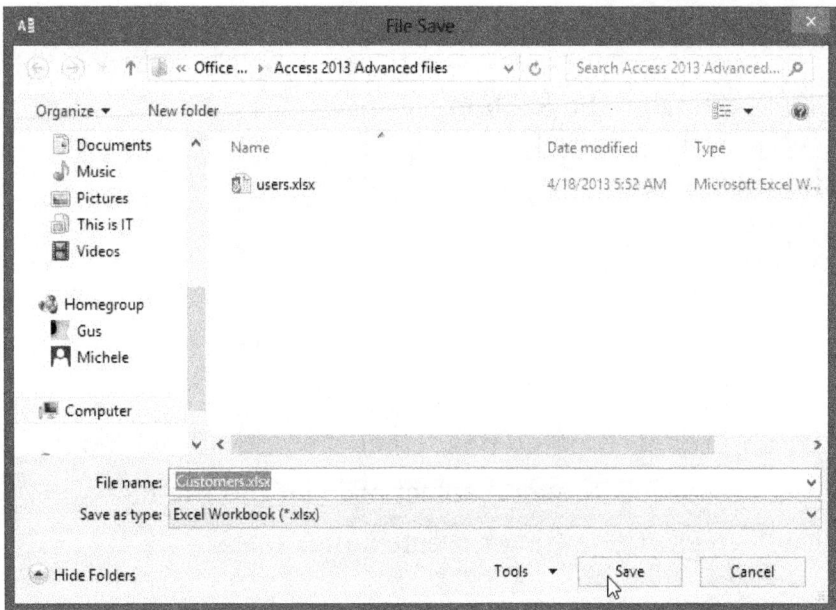

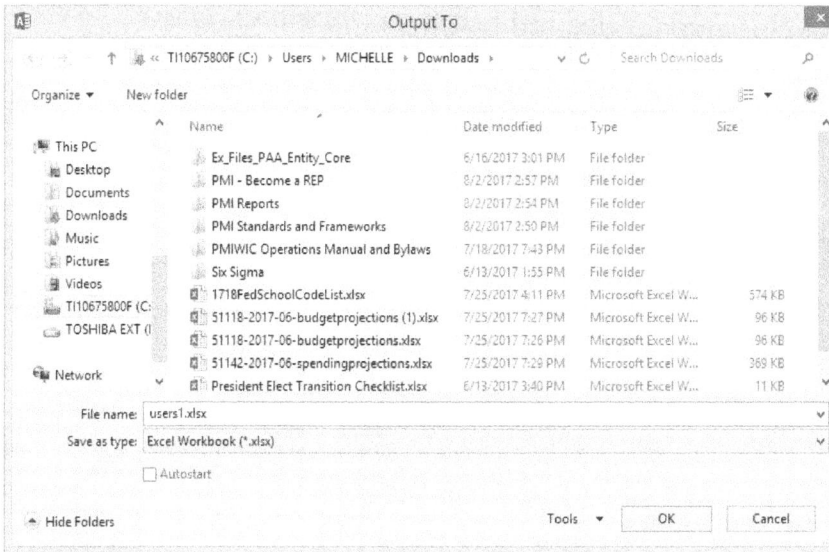

Step 5: Select the File Format from the drop-down list.

Step 6: Check the Export data with formatting and layout box if you want to keep most of the layout and formatting information.

Step 7: If you checked the first box, you can check the Open the destination file box to see the results of your export as soon as it is finished.

Step 8: If you selected specific records before choosing the Export command, you can check or clear the Export only the selected records box.

Step 9: Select OK.

Step 10: You can save the export steps if desired by checking the Save Export Steps box. Select Close.

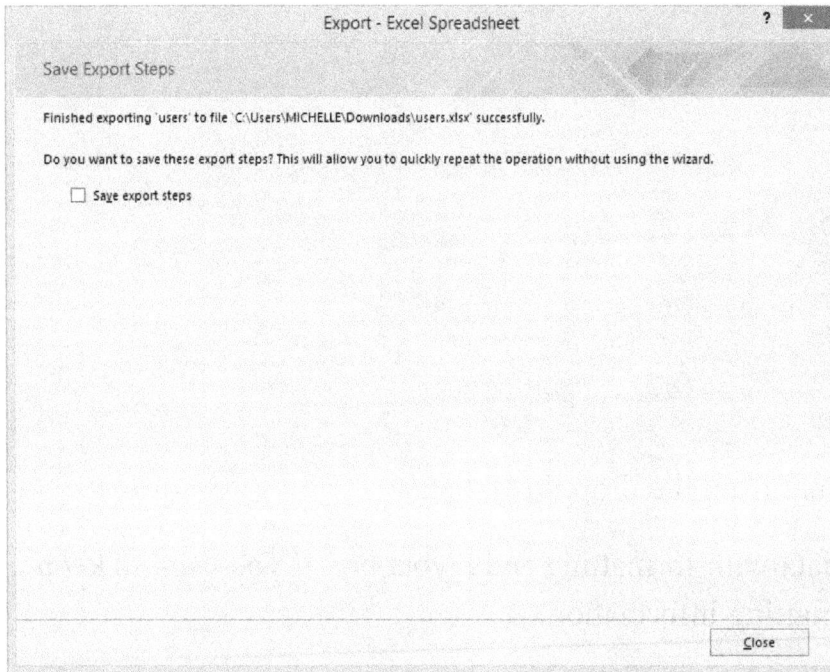

Chapter 14: Creating Queries

In this chapter, you will learn how to create queries without using the Wizard. We will start with the select query. You will also learn how to create a make table query, an append query, and a cross tab query. You will also learn how to modify the tables included in your queries by using the Show or Remove tables commands.

Creating a Select Query

The Select Query is the most basic type of query, and the starting point for many of the other types of queries.

To create a select query, use the following procedure.

Step 1: Open the database that contains the records you want to select.

Step 2: Select the Create tab from the Ribbon.

Step 3: Select Query Design.

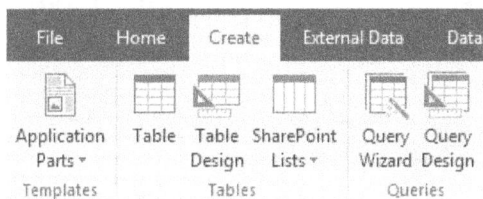

Step 4: Access displays the Show Table dialog box. Double-click the table(s) or queries that contain the records you want to copy. Or highlight them in the list and select Add.

Step 5 The tables or queries appear in the Query designer window on the top half of the window.

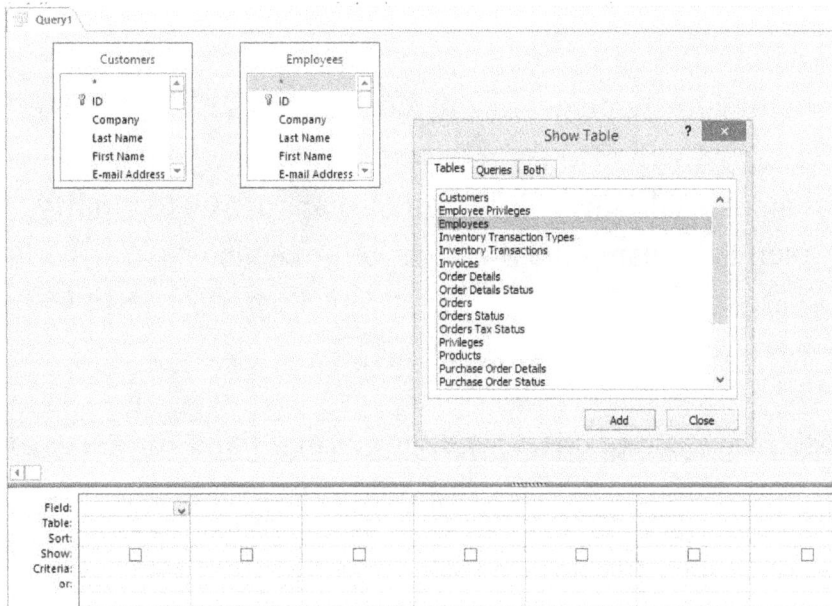

Step 6: Select Close to close the Show Table dialog box.

Step 7: From the tables in the top half of the designer window, double-click each field that you want to include in the query. Or you can double-click the asterisk at the top to quickly add all fields in the table. The selected fields appear in the Field row on the query design grid on the bottom half of the window.

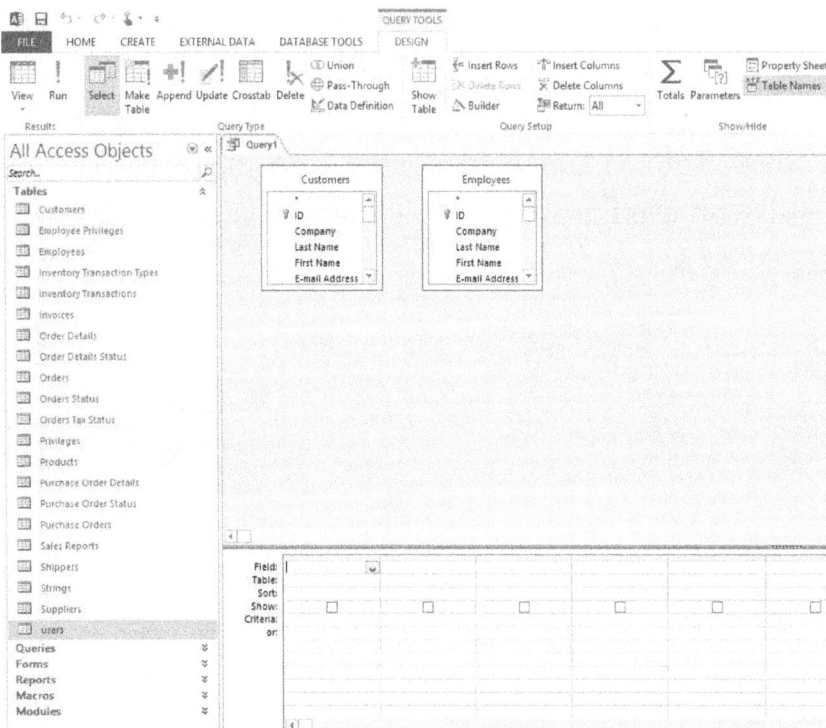

Step 8: Select the Design tab from the Query Tools Ribbon.

Step 9: Select Run. This is another way of executing the query from Query Design view.

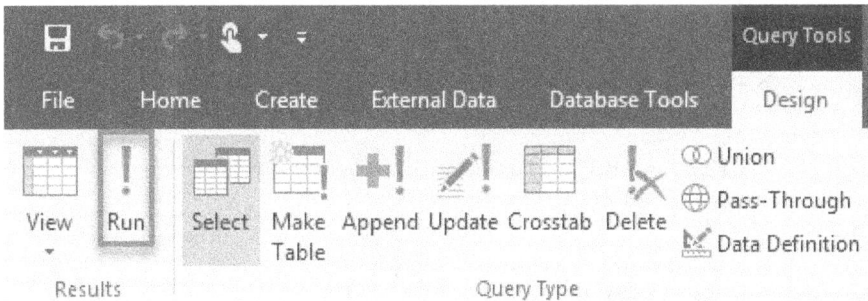

Step 10: Verify the results of the query. You can add or remove fields by switching back to Query Design view.

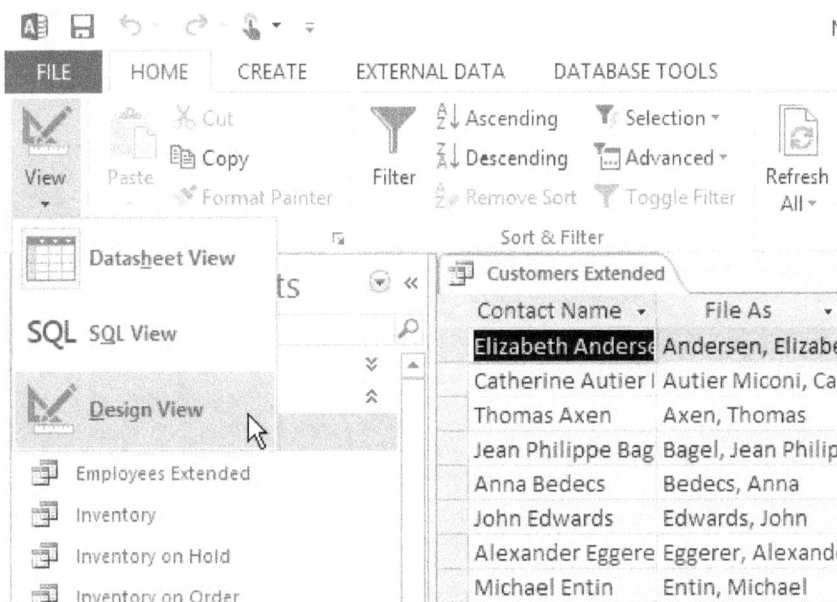

Step 11: Save the query by selecting the save icon. Enter a name for the query that is not the same as another table or query. Select OK.

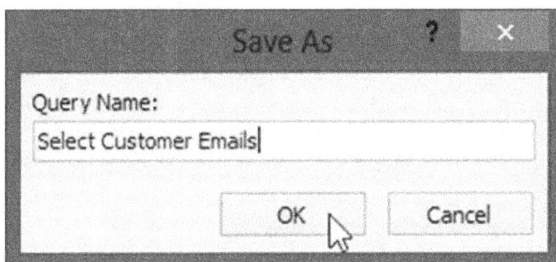

Creating a Make Table Query

The Make Table query is the first action query we will discuss. You can use a make table query to copy or archive the information in a table.

To create and run a make table query, use the following procedure.

Step 1: Open the select query that you want to use in Design view.

Step 2: Select the Design tab from the Query Tools Ribbon.

Step 3: Select Make Table.

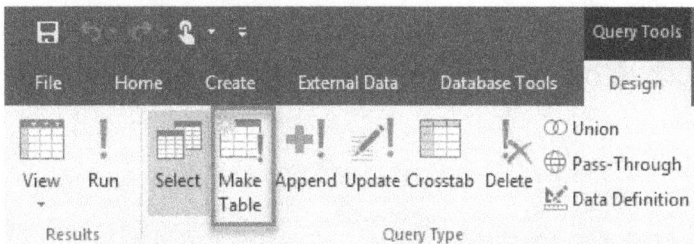

Step 4: In the Make Table dialog box, select from the drop-down list or enter a Table Name for the table that will be created.

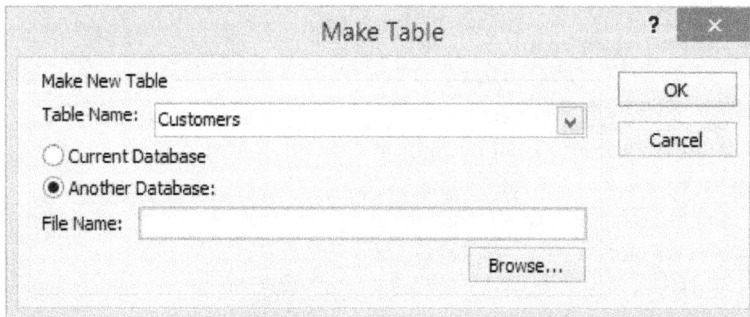

Step 5: Select Another Database to create the table in another database. Select Browse to navigate to the location of that database and select OK.

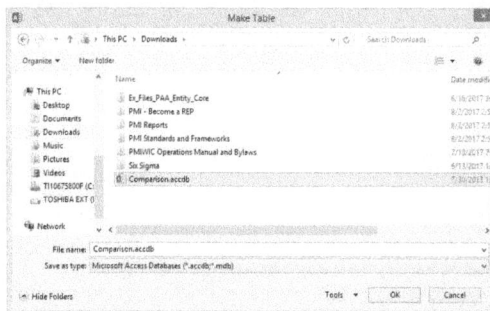

Step 6: On the Design tab on the Query Tools Ribbon, select Run to execute the query.

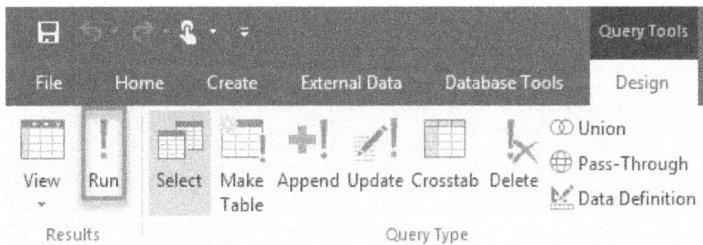

Access displays a confirmation message. Select Yes to continue.

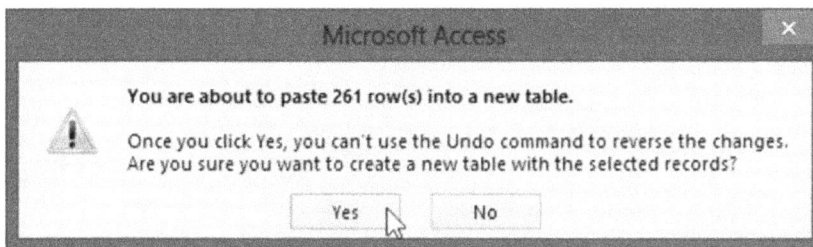

Step 7: Now open the destination database to see the results of the new table.

Creating an Append Query

An append query adds new records to an existing table.

To create another, select query to select the employees this time in the Northwind sample database, use the following procedure.

Step 1: Open the database that contains the records you want to select.

Step 2: Select the Create tab from the Ribbon.

Step 3: Select Query Design.

Step 4: Double-click the table(s) or queries that contain the records you want to copy.

Step 5: Select Close to close the Show Table dialog box. The tables or queries appear in the Query designer window on the top half of the window.

Step 6: Double-click each field that you want to include. Or you can double-click the asterisk at the top to quickly add all fields. The selected fields appear in the Field row on the query design grid on the bottom half of the window.

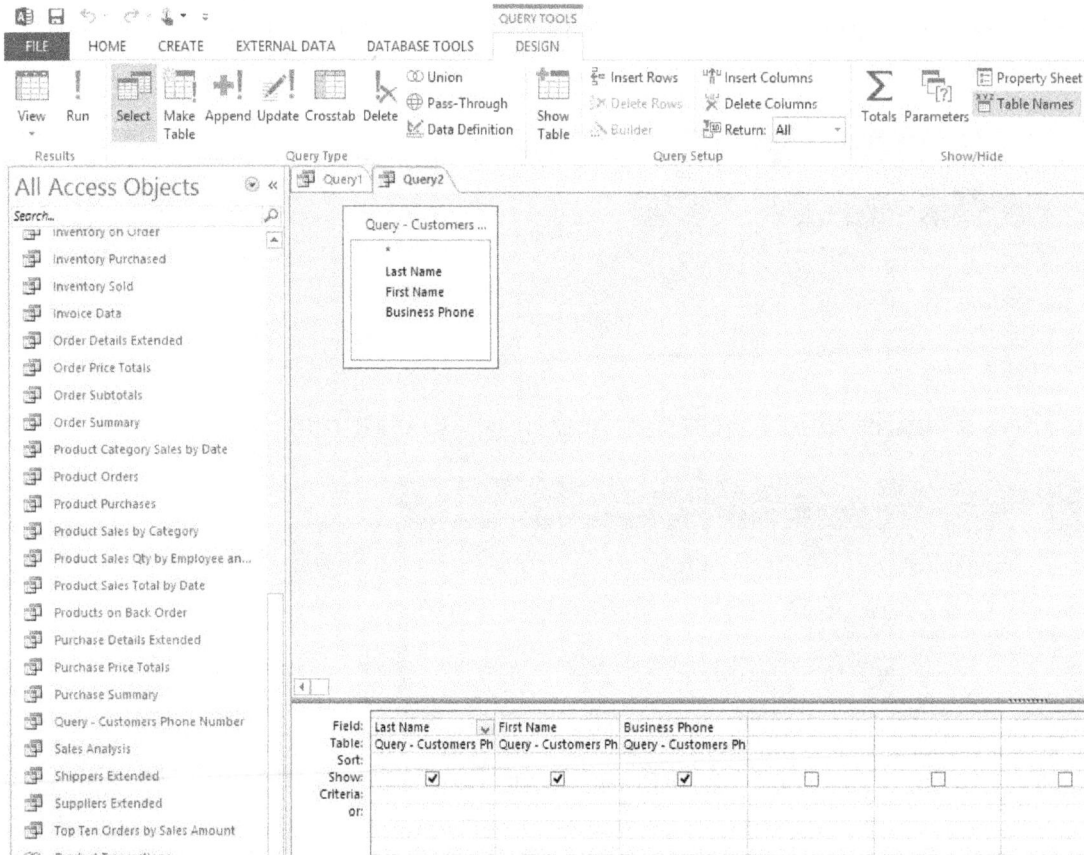

Step 7: Select the Design tab from the Query Tools Ribbon.

Step 8: Select Run.

Step 9: Verify the results of the query. You can add or remove fields by switching back to Query Design view.

Step 10: Save the query by selecting the save icon. Enter a name for the query that is not the same as another table or query. Select OK.

To convert the select query to an append query.

Step 1: Return to Design View by selecting View from the Home tab on the Ribbon. Select Design View.

Step 2: Select the Design tab from the Query Tools Ribbon.

Step 3: Select Append.

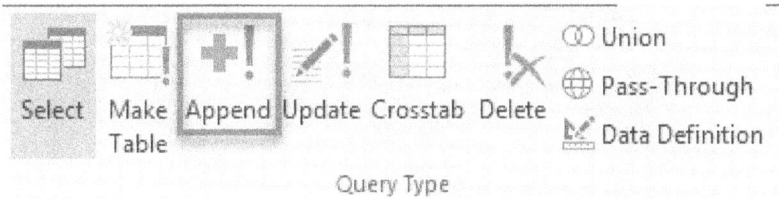

Step 4: In the Append dialog box, select Another Database. Select Browse to navigate to the location of that database. Select OK.

Step 5: Select the Table Name from the drop-down list.

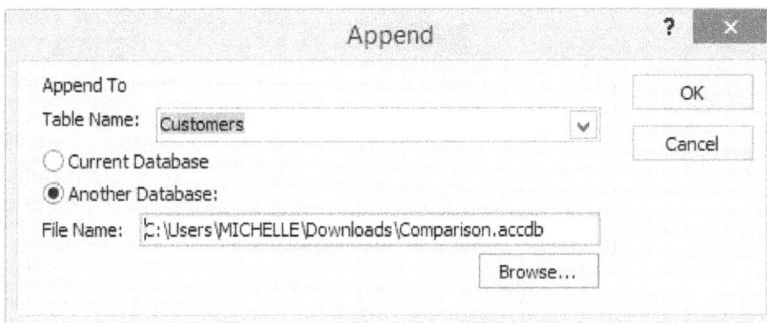

Step 6: Select OK.

Depending on how you created your select query, the appropriate fields are automatically added. Unmatched fields are left blank. You can select a new option from the Append To drop down list at the bottom if necessary. If the Append To (destination field) is left blank, Access will not append data to that field. The destination table name is at the top of the Append To list.

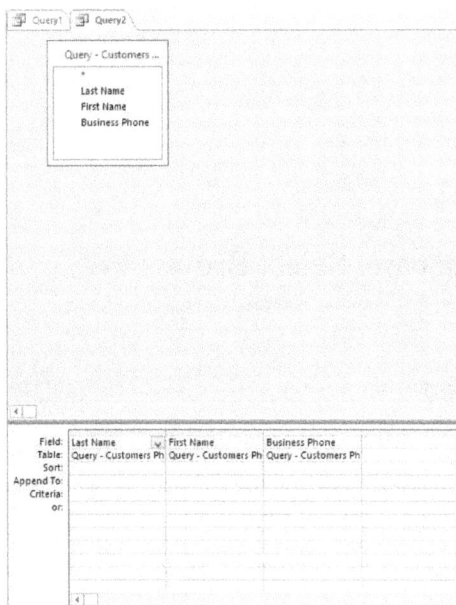

To preview and run the append query, use the following procedure.

Step 1: Switch to Datasheet view to preview your query results.

Step 2: Switch back to Design view.

Step 3: Select Run.

Step 4: If you receive a warning message, select Yes to continue.

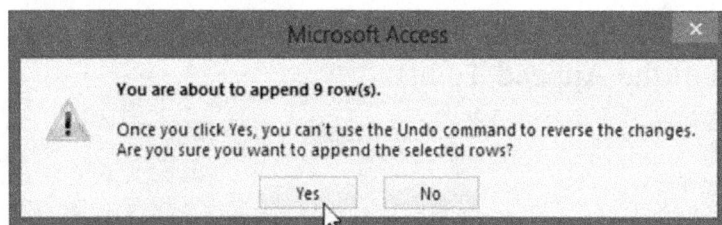

Step 5: Now open the destination database to see the results of the new table.

Creating a Cross Tab Query

A cross tab query allows you to summarize data in a compact format like a spreadsheet.

To create a cross tab query, use the following procedure.

Step 1: Create a select query using the Invoices table in the sample database. The query should include the following fields:

- Quarter
- Customer
- Project
- Invoice Amount

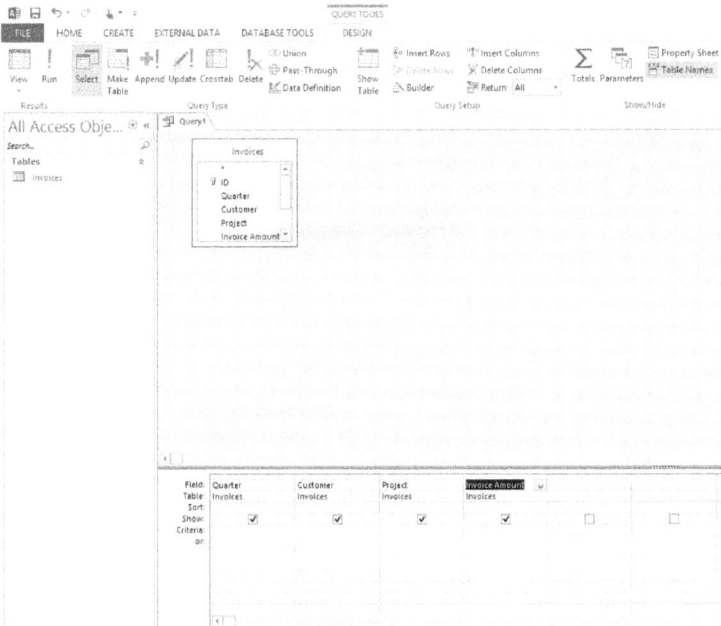

Step 2: Now select Crosstab from the Design tab on the Query Tools Ribbon.

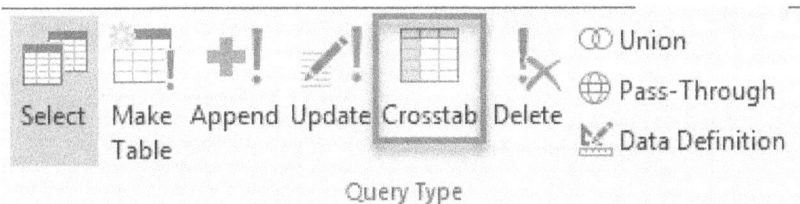

Step 3: In the bottom section, we will need to define a row heading, a column heading, and a value. Let's make the Quarter field the column heading, the Customer the Row heading, and the Invoice Amount the value.

- Select Column Heading from the Crosstab drop down list under Quarter.
- Select Row Heading from the Crosstab drop down list under Customer.
- Select Value from the Crosstab drop down list under Invoice Amount. Also under Invoice Amount, we are going to change the Total from Group by to Sum.
- For the Project column, we will set the Total as Sum and the Crosstab setting as (not shown).

Field:	Quarter	Customer	Project	Invoice Amount
Table:	Invoices	Invoices	Invoices	Invoices
Total:	Group By	Group By	Sum	Sum
Crosstab:	Column Heading	Row Heading	(not shown)	Value
Sort:				Row Heading
Criteria:				Column Heading
or:				**Value**
				(not shown)

Field:	Quarter	Customer	Project	Invoice Amount
Table:	Invoices	Invoices	Invoices	Invoices
Total:	Group By	Group By	Sum	Sum
Crosstab:	Column Heading	Row Heading	(not shown)	Group By
Sort:				**Sum**
Criteria:				Avg
or:				Min
				Max
				Count
				StDev
				Var
				First
				Last
				Expression
				Where

Step 4: Select Run.

Step 5: Now view the results of your query.

Customer	<>	Q1	Q2	Q3	Q4
	$96,445.00				
Ad. Lang, Provi			$1,429.00		
Business Softw		$3,059.00			
Celpian					$5,535.00
Centro Latino		$2,015.00			
CTM Solutions		$9,144.00	$5,127.00	$2,269.00	
Decker Rights,					$2,341.00
Docket Trainin					$3,178.00
Documents Are			$5,671.00		
Educational Inf					$9,584.00
EFI Networks T			$3,446.00		$2,027.00
Media Learnin;		$1,367.00			
Milennium Tra				$10,917.00	$10,961.00
PPI, Inc.		$6,714.00			
ProSource Disc			$1,781.00		
ProSource Trali		$3,443.00	$3,977.00		
Wings Source			$1,800.00		
Zodiac Informa					$1,614.00

Showing and Removing Tables

Showing and Removing Tables will help when you need to modify an existing query.

120

To show the Show Table dialog box, use the following procedure.

Step 1: Open the query that you want to modify.

Step 2: Select the Design tab from the Query Tools Ribbon.

Step 3: Select Show Table.

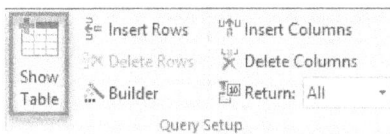

Step 4: Remember that in the Show Table dialog box, you can show Tables, Queries, or a list of both. You can hold down the SHIFT or CTRL keys to select multiple tables to add to the Query Design before selecting Add.

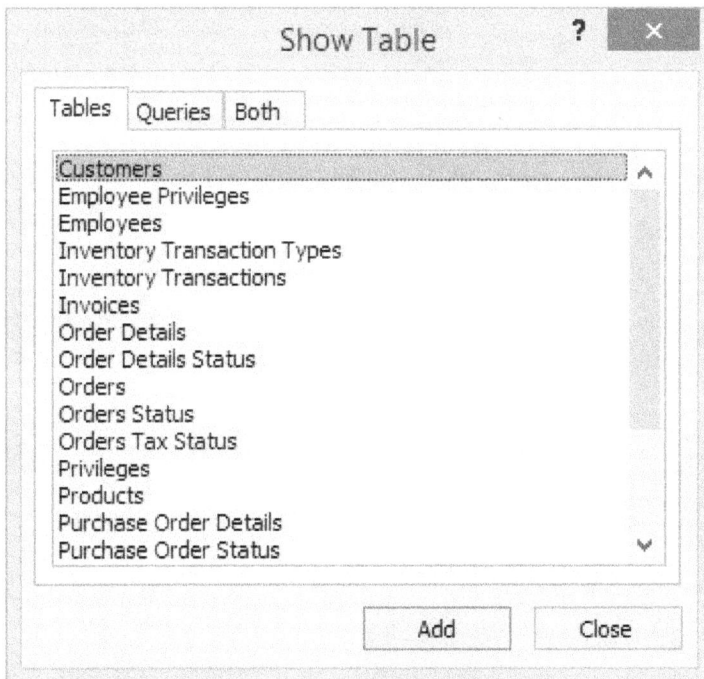

To remove a table from the query design, use the following procedure.

Step 1: Right click on the table that you want to remove in the Query Design window.

Step 2: Select Remove Table from the context menu. Note that all fields in the query that originated from that table are also removed in the bottom section.

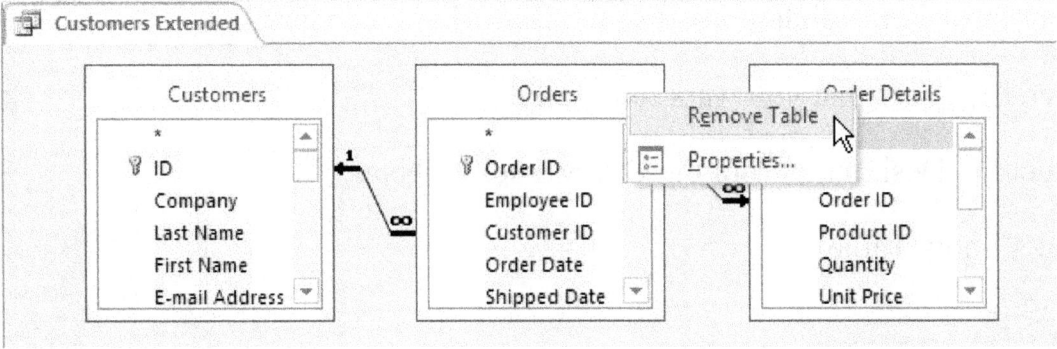

Chapter 15: Creating Forms and Reports

In this chapter, you will start creating forms and reports. First, you will use the form wizard to create a form. You will also use the report wizard to create a report. Then you will look at the tools available in both the form layout view and the report layout view.

Creating a Form with the Form Wizard

The Form Wizard helps you to develop forms from selected data.

This example uses the Customer table as a starting point, use the following procedure.

Step 1: In the Navigation pane, highlight the table (or query) that you want to use on your form.

Step 2: Select the Create tab on the Ribbon.

Step 3: Select Form Wizard.

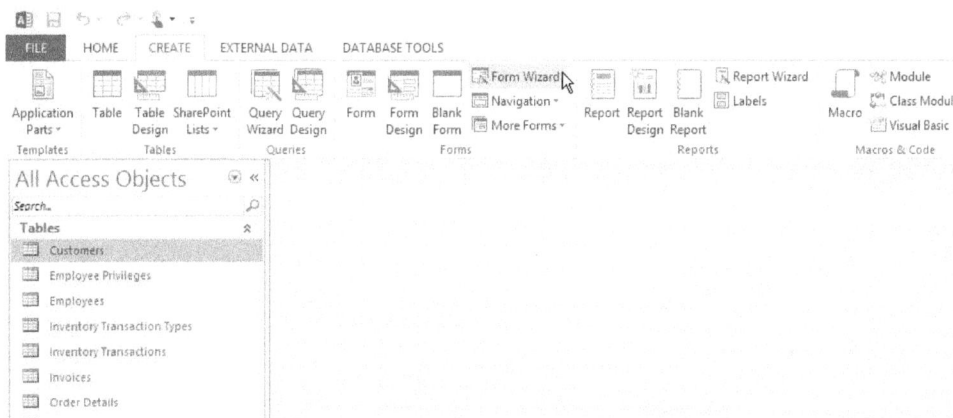

Access opens the Form Wizard.

Step 1: You can select more than one table or query for the data you want on your form. The table you highlighted in the Navigation pane is selected, but you can change it by selecting a new item from the Tables/Queries drop down list.

Step 2: The fields available on the selected table appear in the Available Fields column. Double-click the fields you want on your form, or highlight the field(s) and select the right arrow (or the double right arrow to select all). The items in the

Selected Fields column will appear on your form. To remove an item from the Selected Fields column, highlight it and select the left arrow (or the double left arrow to remove all). To add fields from an additional table, return to step 4.

Step 3: When you have finished selecting the fields to appear on your form, select Next.

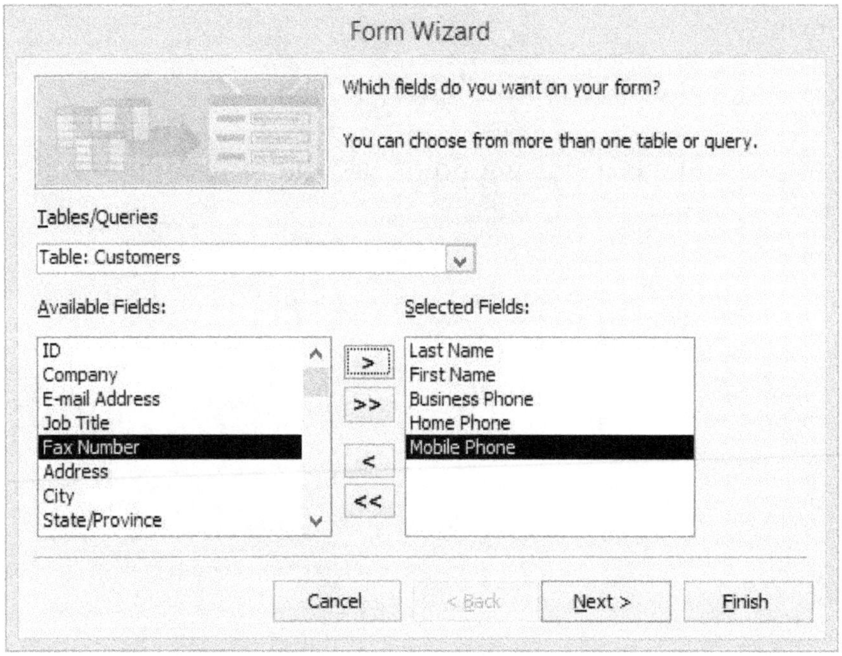

The next screen on the wizard allows you to select a standard layout for your form.

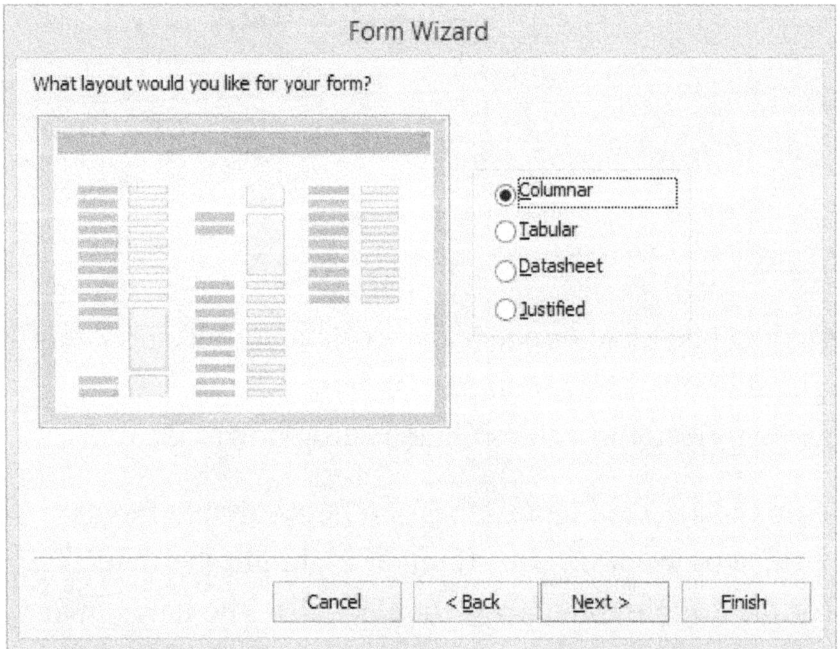

Step 4: Select one of the layout options and select Next.

The final screen of the wizard allows you to name your form.

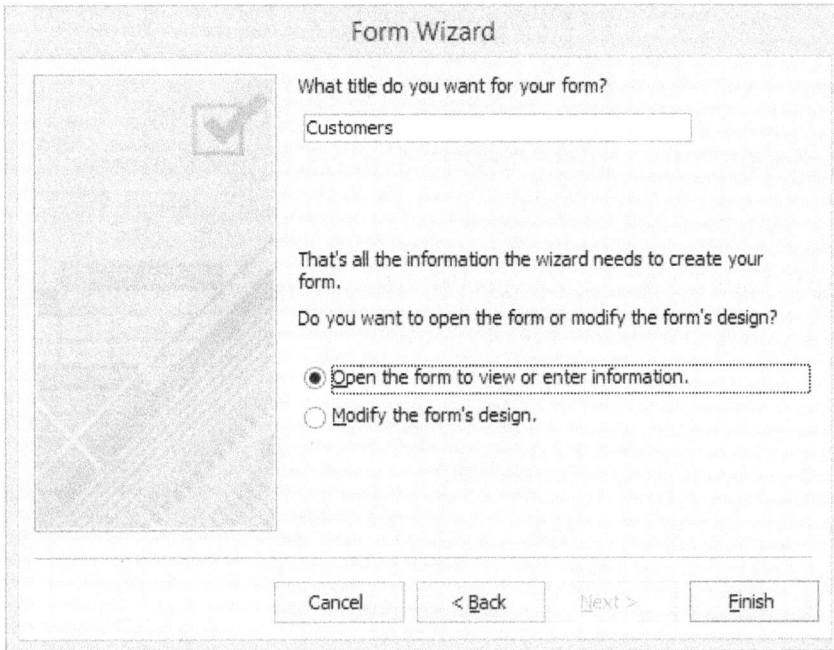

Step 5: Verify the default name, or enter a new name for the form.

Step 6: Select whether you want to open the form to view or enter information, or if you want to modify the form's design.

Step 7: Select Finish.

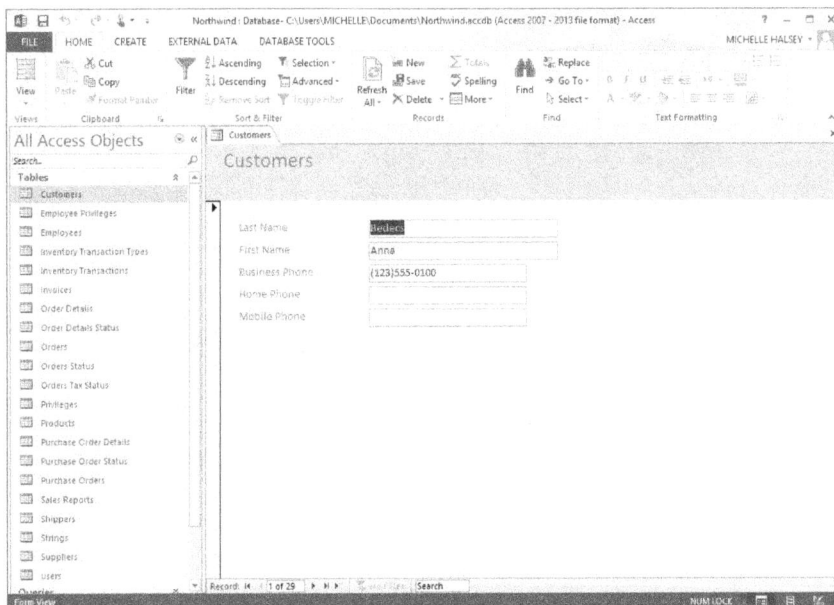

Creating a Report with the Report Wizard

The Report Wizard is like other wizards in Access used to help you create objects.

To create a report using the Report Wizard.

Step 1: In the Navigation pane, highlight the table (or query) data that you want to use in your report.

Step 2: Select Report Wizard from the Create tab on the Ribbon.

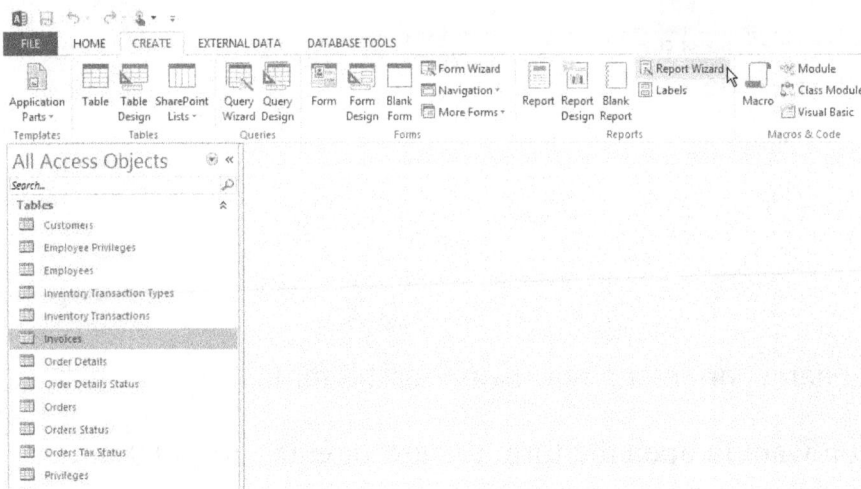

Access opens the Report Wizard.

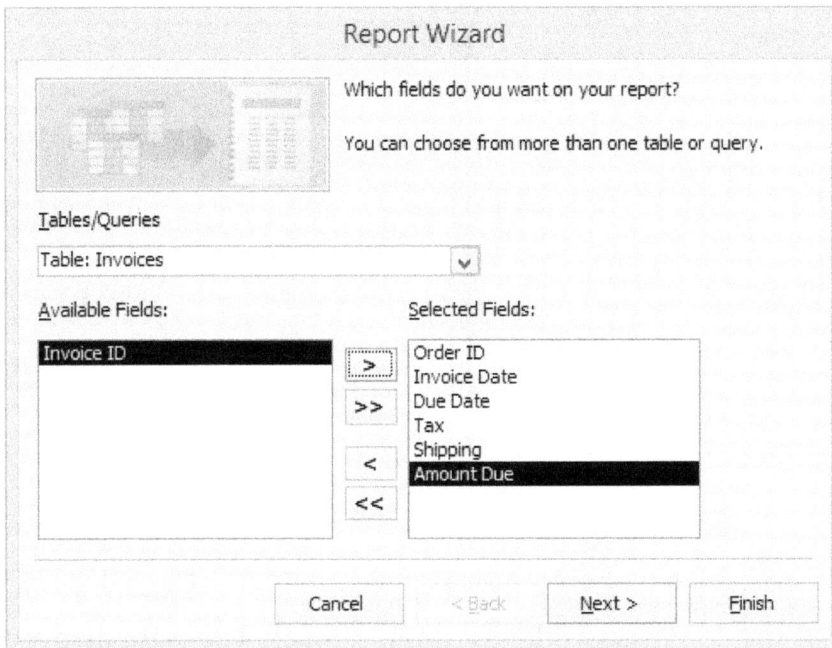

Step 3: You can select more than one table or query for the data you want on your report. The table you highlighted in the Navigation pane is selected, but you can change it by selecting a new item from the Tables/Queries drop down list.

Step 4: The fields available on the selected table appear in the Available Fields column. Double-click the fields you want on your report, or highlight the field(s) and select the right arrow (or the double right arrow to select all). The items in the Selected Fields column will appear on your report. To remove an item from the Selected Fields column, highlight it and select the left arrow (or the double left arrow to remove all). To add fields from an additional table, return to step 3.

Step 4: When you have finished selecting the fields to appear on your report, select Next.

The next screen on the wizard allows you to group the information on your report. Groups allow you to separate records visually. A group includes introductory and summary information, the detail records, and a footer.

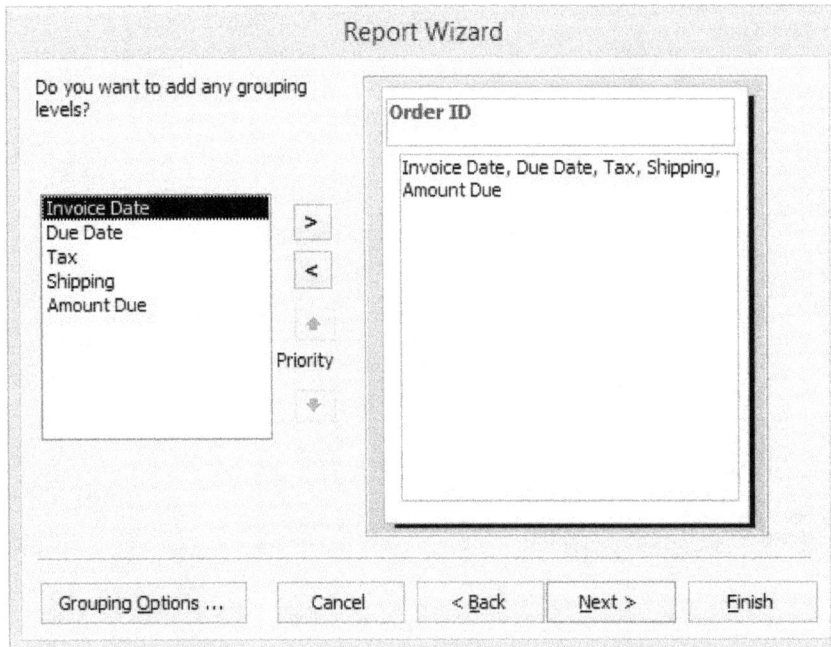

Report Wizard

Do you want to add any grouping levels?

Order ID

Invoice Date, Due Date, Tax, Shipping, Amount Due

Invoice Date
Due Date
Tax
Shipping
Amount Due

>
<

Priority

Grouping Options ...   Cancel   < Back   Next >   Finish

Step 5: Select one or more of the controls to add a grouping level. The right arrow adds the grouping level. The left arrow removes a grouping level. The Priority arrows allow you to rearrange the levels if you have selected more than one control for grouping.

Step 6: The Grouping Options button opens a new window, which differs, depending upon the type of control for the selected grouping option. A sample is illustrated below.

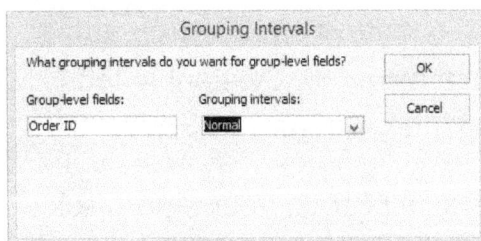

Grouping Intervals

What grouping intervals do you want for group-level fields?

Group-level fields: Grouping intervals:

Order ID   Normal

OK   Cancel

Step 7: Select an appropriate Grouping interval from the drop-down list and select OK.

Step 8: Select Next on the Report Wizard.

The next screen on the wizard allows you to choose the sort order for your data.

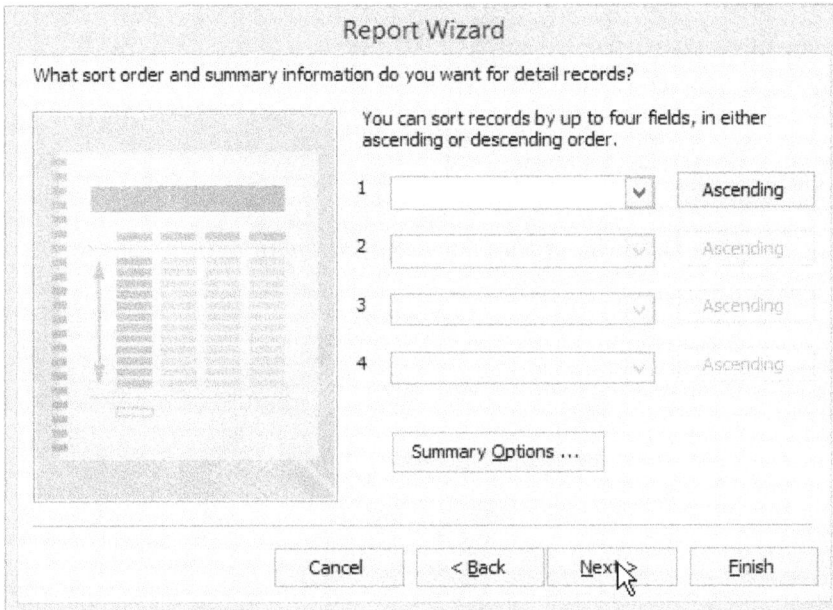

Step 9: Select up to four fields for sorting and select either Ascending or Descending for the order.

Step 10: The Summary Options button opens a new window, which differs, depending upon the type of controls on the report. A sample is illustrated below.

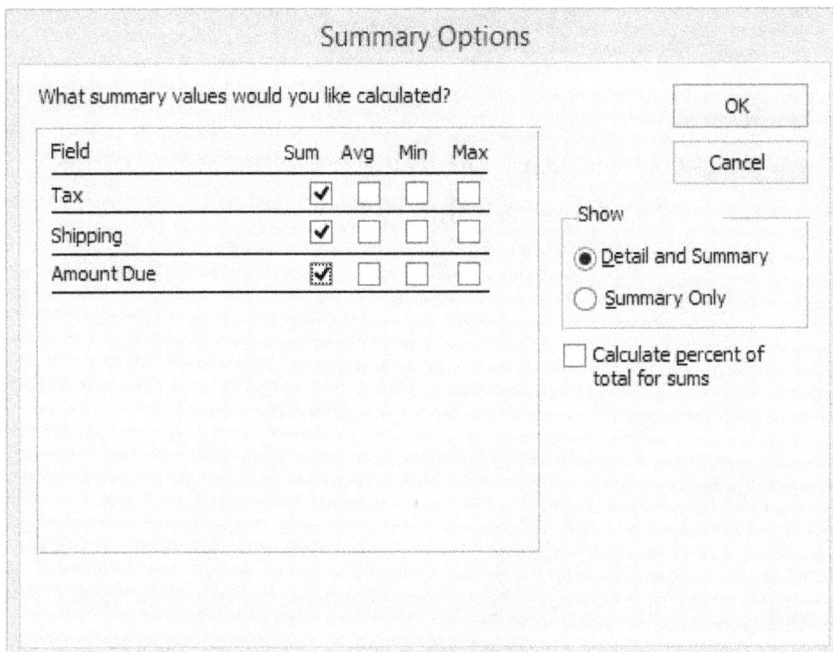

Step 11: You can choose Summary, Average, Minimum, or Maximum values to be calculated. You can choose whether to show detail and summary or just summary. You can also calculate a percent of total for all sums. When you have finished setting the summary options, select OK.

Step 12: Select Next on the Report Wizard.

The next screen on the wizard allows you to select from a list of layout and orientation options.

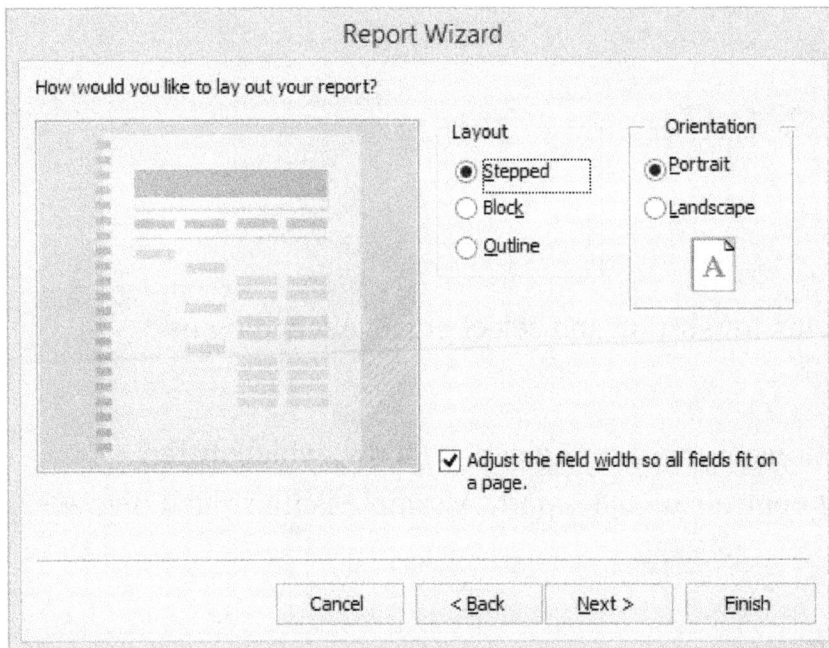

Step 13: Select a layout option and an Orientation option. Check the Adjust the field width so all fields fit on a page box if desired. Select Next.

The final screen of the wizard allows you to name your report.

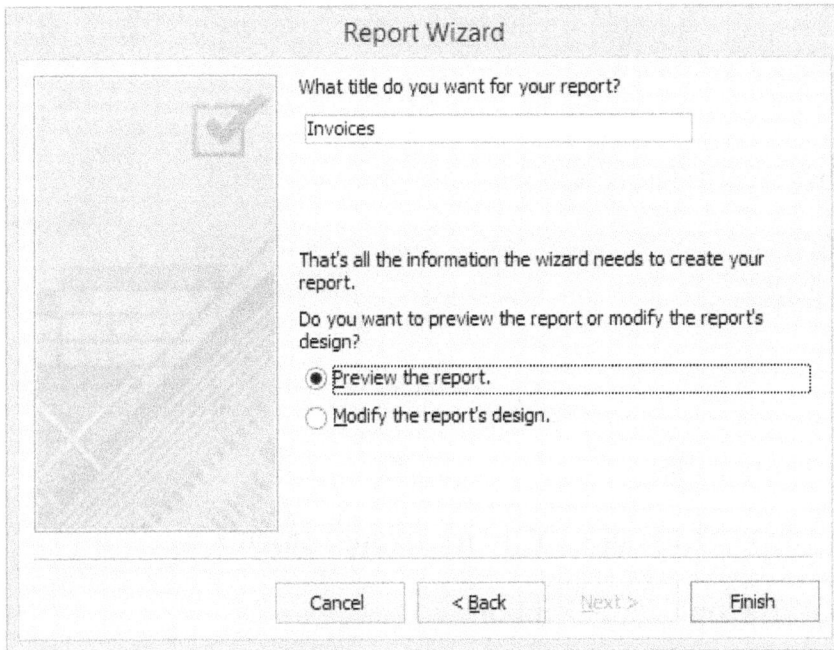

Step 14: Verify the default name, or enter a new name for the report.

Step 15: Select whether you want to open the report, or if you want to modify the report's design.

Step 16: Select Finish.

Modifying Form Layout

In Layout view, there are three tabs with commands to help you customize your form.

The Design tab on the Form Layout Tools Ribbon.

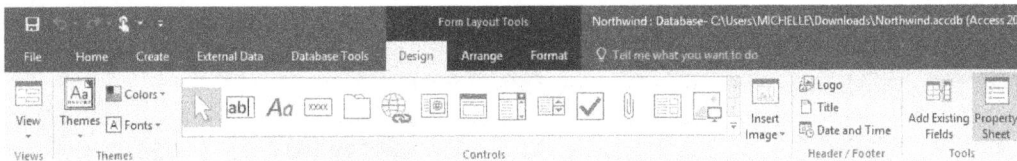

The Arrange tab on the Form Layout Tools Ribbon.

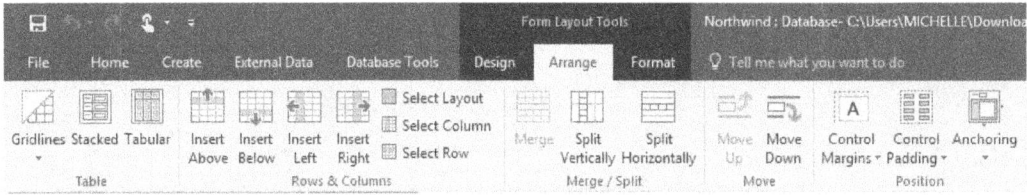

The Format tab on the Form Layout Tools Ribbon.

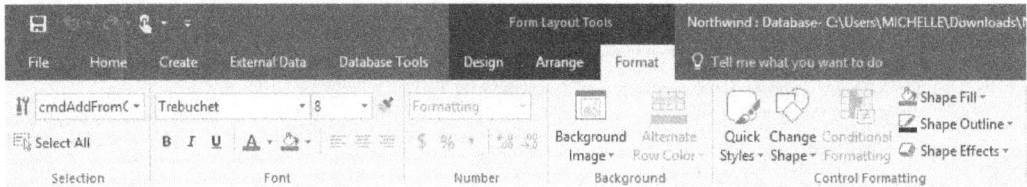

Key Features on the Report Tools Tabs

Your reports also can be opened in Layout view to make adjustments.

The Design tab on the Report Layout Tools Ribbon.

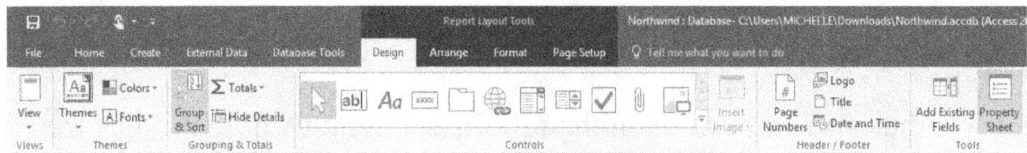

The Arrange tab on the Report Layout Tools Ribbon.

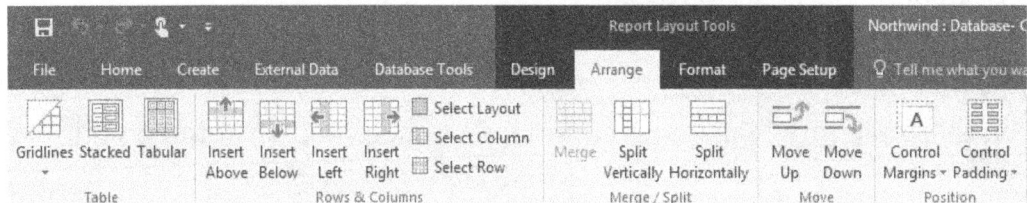

The Format tab on the Report Layout Tools Ribbon.

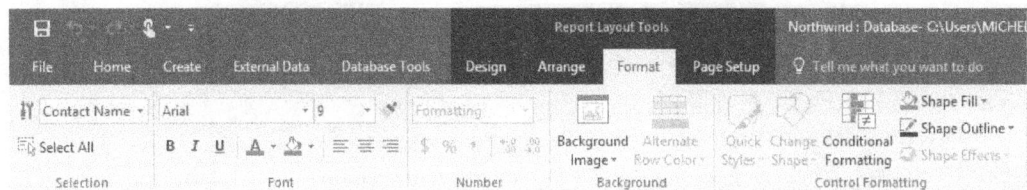

The Page Setup tab on the Report Layout Tools Ribbon.

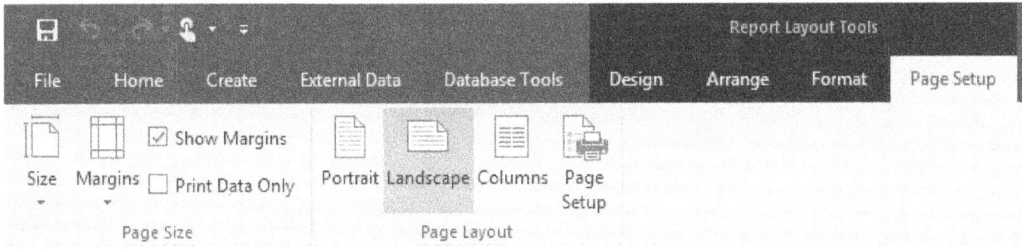

Chapter 16: Working with Validation Rules and Messages

Now we will move to working in Desktop Databases. Although many of the skills you learned when working in apps applies to desktop databases (and in some cases, vice versa), not all desktop options are available in apps. The skills in this chapter are a prime example. We will look at restricting data input with validation rules and messages in this chapter.

About Restricting Data Input

This lesson explains the ways you can make sure that your database users only enter certain types of data.

There are several ways of restricting data input. The data type you select for a field creates one type of restriction, because for example, date/time fields only accept dates and times, and currency fields only accept monetary data.

You can also restrict data by using field properties. For example, if you have set a field size, the user cannot enter data longer than that size.

Another way of restricting input is through Input Masks, which forces users to enter values in a specific way, such as forcing dates in a European format, such as 2016.04.28.

In this chapter, we will focus on a fourth way to restrict input – validation rules. A validation rule is one way to restrict input in a table field or control on a form. Validation text lets you provide a message to help users understand the requirements of the field.

These different methods can be used alone, or in combination with each other.

There are two basic types of validation rules:

- Field Validation Rule
- Record validation Rule

When a user inputs data where there is a validation rule, Access checks whether the input breaks the rule. If the rule is broken, Access does not accept the input and displays a message.

Creating a Field Validation Rule

You can use a validation rule to specify a criterion that all valid field values must meet.

To review the existing field validation rule in the sample database, use the following procedure.

Step 1: Open the Orders table in Design view.

Step 2: Select the Order Date field.

Step 3: View the Validation Rule and Validation Text.

To create a field validation rule in Datasheet view, use the following procedure.

Step 1: Open the table with the field you want to control in Datasheet view.

Step 2: Select the field you want to validate.

Step 3: Select the Fields tab from the Table Tools Ribbon.

Step 4: Select Validation.

Step 5: Select Field Validation Rule.

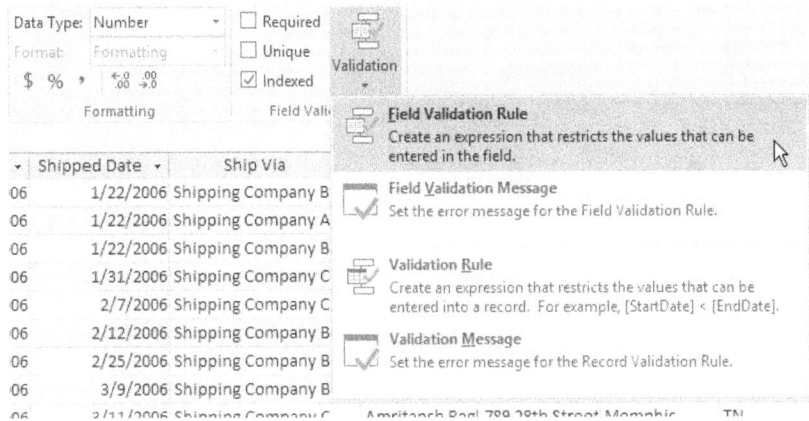

Step 6: Access opens the Expression Builder to help you enter your expression. Enter your expression or use the area at the bottom to help you build your expression. When you have finished, select OK.

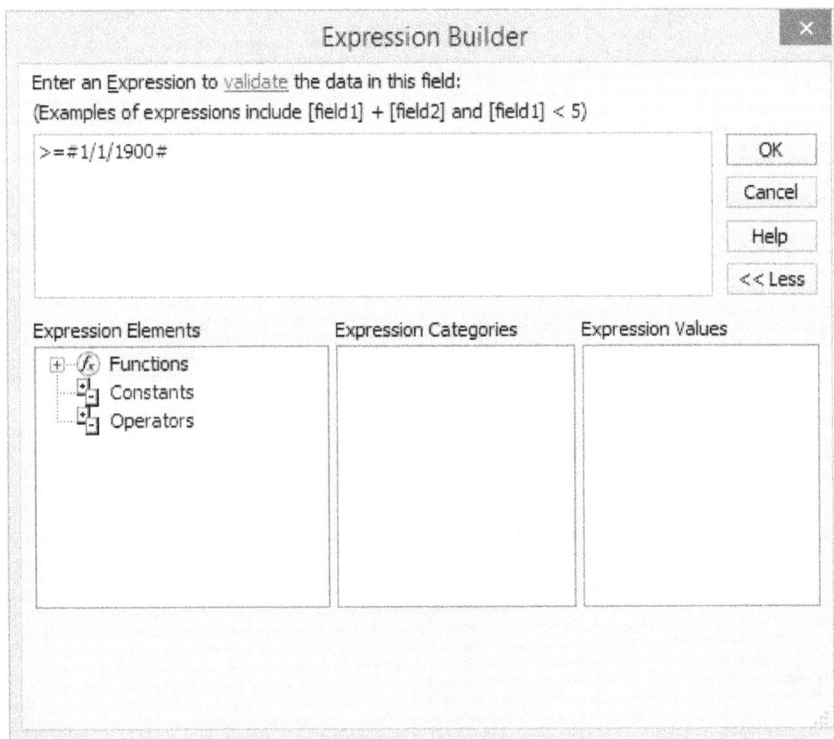

Step 7: To create a message for users that enter incorrect values, select the Validation command from the Table Tools Fields tab again.

Step 8: Select Field Validation Message.

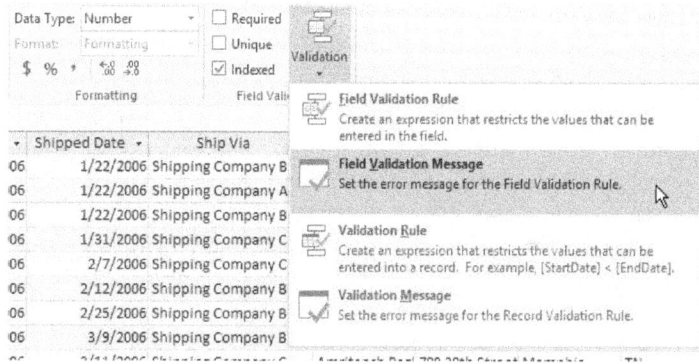

Step 9: In the Enter Validation Message dialog box, enter the text that you want to display. Select OK.

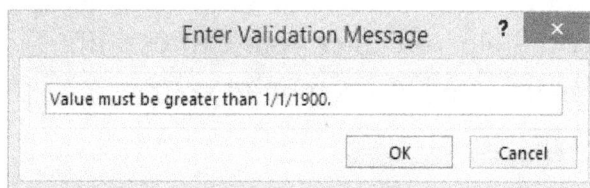

Step 10: Save the changes to your table. Your rule is ready for testing.

Creating a Record Validation Rule

A record validation rule checks input to one or more fields in a record. It is applied when the focus leaves the record.

To create a record validation rule, use the following procedure.

Step 1: Open the table you want to control in Datasheet view. In this example, we will use the Orders table from the sample database.

Step 2: Select the Fields tab from the Table Tools Ribbon.

Step 3: Select Validation.

Step 4: Select Validation Rule.

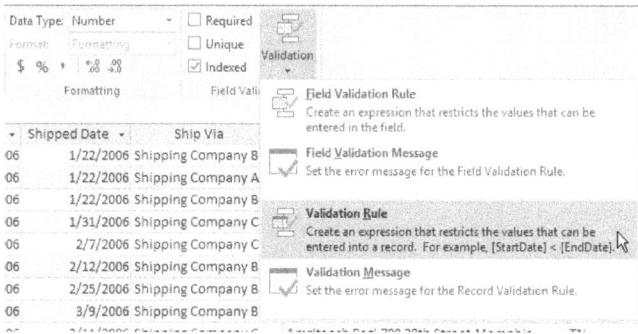

Step 5: Access opens the Expression Builder to help you enter your expression. Enter your expression or use the area at the bottom to help you build your expression. When you have finished, select OK.

In our example, we will enter the following expression:

[Shipped Date]<=[Order Date} + 10

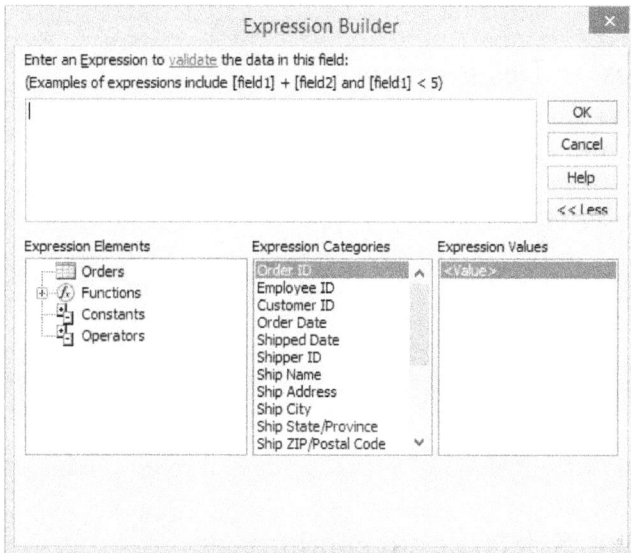

Step 6: To create a message for users that enter incorrect values, select the Validation command from the Table Tools Fields tab again.

Step 7: Select Validation Message.

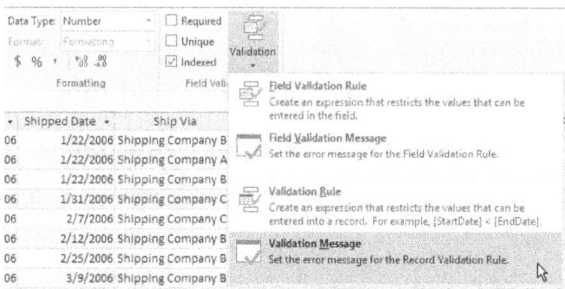

Step 8: In the Enter Validation Message dialog box, enter the text that you want to display. Select OK.

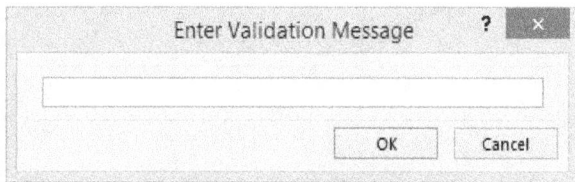

Step 9: Save the changes to your table. Your rule is ready for testing.

Testing Validation Rules

If you add a validation rule to an existing table, you might want to test the rule to see whether any existing data is not valid.

To test a validation rule, use the following procedure.

Step 1: Open the table that you want to test in Design View.

Step 2: Select the Design tab from the Table Tools Ribbon if it is not already showing.

Step 3: Select Test Validation Rules.

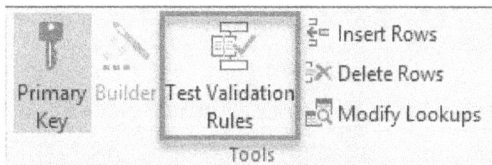

Step 4: Access displays a warning message. Select Yes to continue.

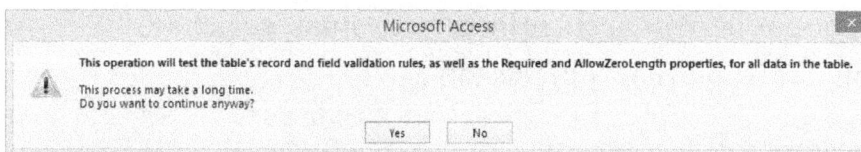

Step 5: If you are prompted to save your table, select Yes.

Step 6: You may see a variety of other alert messages as you proceed. Read the instructions for each message and select Yes or No, as appropriate, to complete or stop the testing.

Microsoft Access

The existing data violates the 'Validation Rule' property for field 'Shipped Date.'

If you continue testing, Microsoft Access will inform you if data violates any other property settings in the table. Would you like to continue testing?

Yes   No

Chapter 17: Working with Macros

This chapter looks at macros in the desktop database environment. We will start with learning how to create a data macro and a named macro. Then, you will learn how to rename and delete macros. The chapter will close with learning about the AutoExec macro.

Creating Data Macros

Creating macros in desktop databases is like creating macros in apps.

To create a data macro from Datasheet view, use the following procedure.

Step 1: Open the table where the data you want to modify is stored.

Step 2: Select the Table tab from the Table Tools Ribbon.

Step 3: Select the type of event where you would like to add the macro. The options are Before Change or Before Delete.

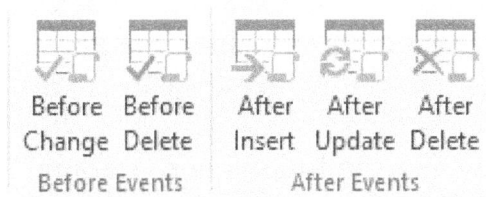

Step 4: In the Macro Design view, select from the drop down list the first action that you want the event to perform. You can also add macro actions from the Action Catalog on the right. Just double-click on the action you want to include. Notice that if you click once on an action in the Action Catalogue, Access displays a definition at the bottom of the Action catalog pane.

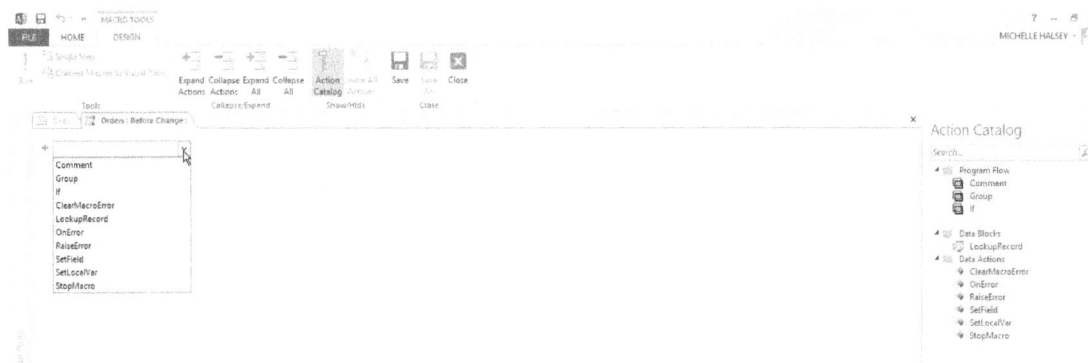

Step 5: If you select actions that require the arguments, fill in the argument information. Note that you can open the Expression Builder in many cases you help you with the arguments.

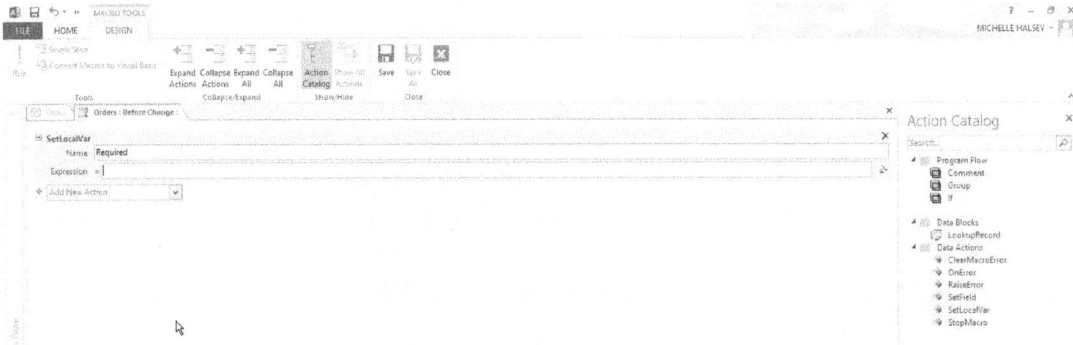

Step 6: If you add multiple actions, you can use the arrow keys on the right to rearrange them. You can also click the X to delete an action you no longer need.

Step 7: Save your changes and close the macro design view.

Creating a Named Macro

This lesson looks at creating data macros from the Design view. Named Macros allow you to create parameters.

To create a named data macro from the Design view, use the following procedure.

Step 1: Open the table where the data you want to modify is stored.

Step 2: Select the Table Tools Design tab from the Table Tools Ribbon.

Step 3: Select Create Data Macros.

Step 4: Select Create Named Macro.

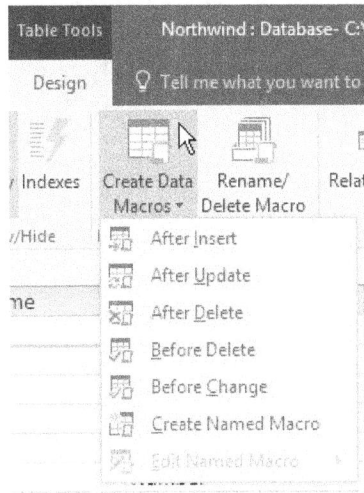

Step 5: In the Macro Design view, select Create Parameter.

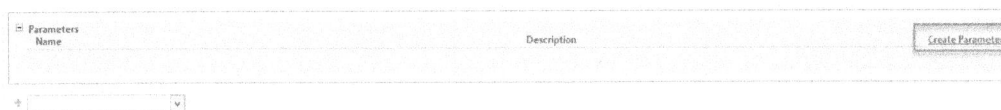

Step 6: In the Name box, type a unique name for the parameter.

Step 7: Enter a Description for the parameter if desired to identify the purpose of the parameter.

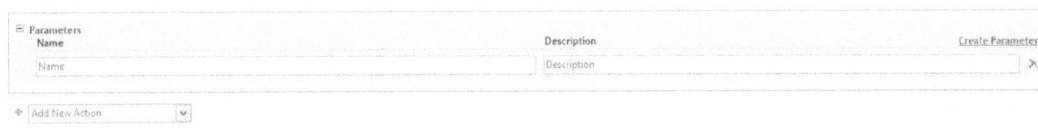

Step 8: Select from the drop down list the first action that you want the event to perform. You can also add macro actions from the Action Catalog on the right. Just double-click on the action you want to include. Notice that if you click once on an action in the Action Catalogue, Access displays a definition at the bottom of the Action catalog pane.

Step 9: If you select actions that require the arguments, fill in the argument information. Note that you can open the Expression Builder in many cases you help you with the arguments.

Step 10: If you add multiple actions, you can use the arrow keys on the right to rearrange them. You can also click the X to delete an action you no longer need.

Step 11: Save your changes. Access displays the Save As dialog box. Enter the name for your macro and select OK.

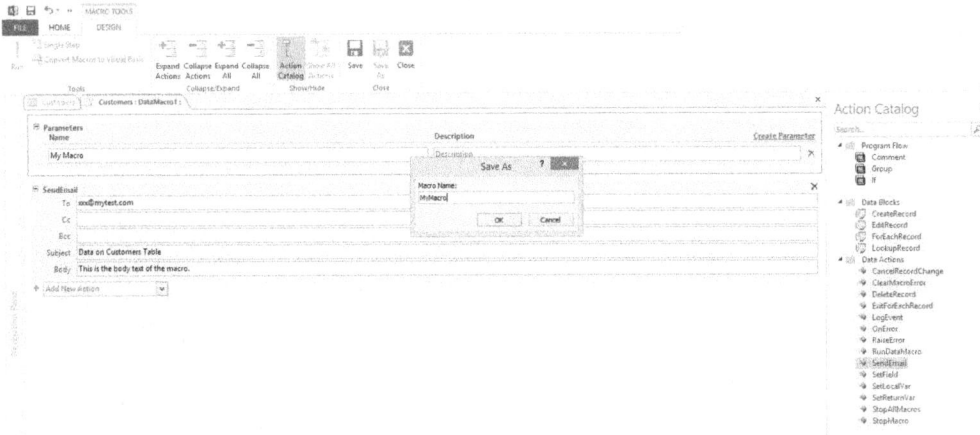

Renaming or Deleting Macros

The Rename/Delete Macro command opens the Data Macro Manager.

To delete or rename macros, use the following procedure.

Step 1: You do not have to have the table with the macro open, but you should have a table open.

Step 2: Select the Table tab (Datasheet view) or the Design tab (Design view) from the Table Tools Ribbon.

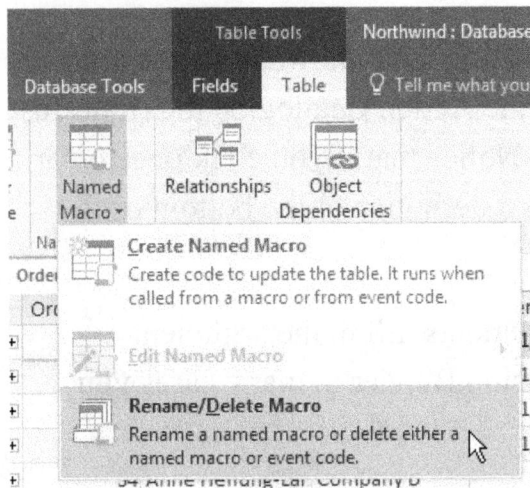

Step 3: If you are in Datasheet View, select Named Macro. Select Rename/Delete Macro. If you are in Design View, select Rename/Delete Macro.

Step 4: The Data Macro Manager dialog box. shows all tables containing data macros. You can use the buttons to the left of the table name to expand or collapse the view.

To rename a macro, select the Rename link to the right of the macro name. The macro name is highlighted. Enter the new name and press Enter.

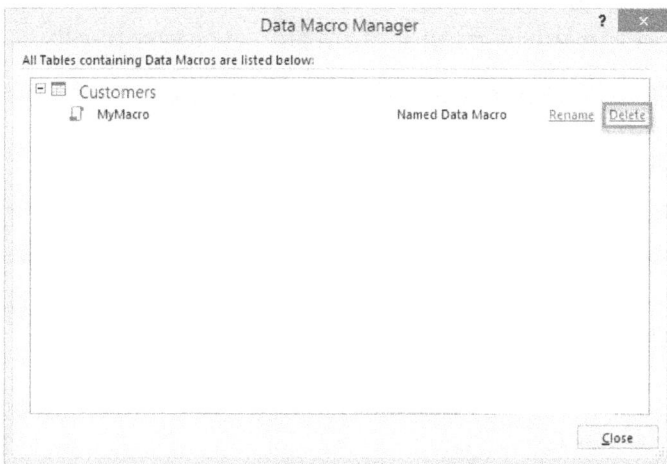

To Delete a macro, select the Delete link to the right of the macro name. Access displays a warning message to confirm the deletion. Select Yes to continue.

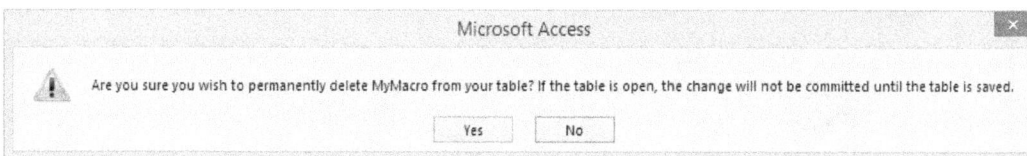

Step 5: Select Close to close the Data Macro Manager.

Step 6: Save the affected table(s).

Creating an AutoExec Macro

If you want to perform a set of actions every time that a database starts, you can create an AutoExec macro.

To create an AutoExec macro, use the following procedure.

Step 1: Select the Create tab from the Ribbon.

Step 2: Select Macro.

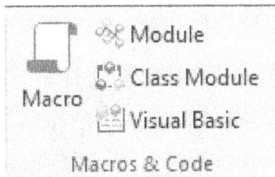

Step 3: In the Macro Design view, select from the drop down list the first action that you want the macro to perform. You can also add macro actions from the Action Catalog on the right. Just double-click on the action you want to include. Notice that if you click once on an action in the Action Catalogue, Access displays a definition at the bottom of the Action catalog pane.

Step 4: If you select actions that require the arguments, fill in the argument information. Note that you can open the Expression Builder in many cases you help you with the arguments.

Step 5: If you add multiple actions, you can use the arrow keys on the right to rearrange them. You can also click the X to delete an action you no longer need.

Step 6: Select Save to save your macro.

Step 7: In the Save As dialog box, enter AutoExec and select OK.

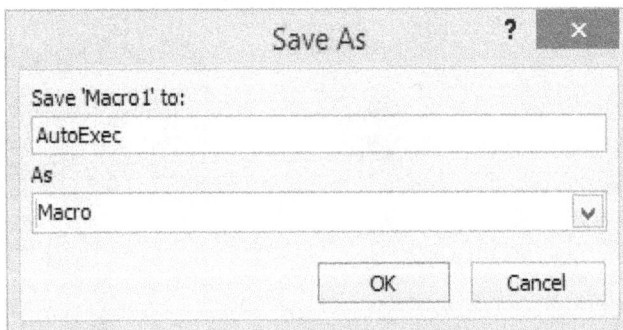

Chapter 18: Advanced Query Tasks

This chapter explains how to create an update query and a parameter query. You will also learn how to use joins to compare the information in two different tables.

Creating an Update Query

An update query can help you update or change existing data in a set of records.

Restrictions on Update Query Capabilities

An update query cannot be used to update data in the following types of fields:

- Calculated fields, because the values in calculated fields do not have a permanent storage location in the table.
- Fields from a totals query or a crosstab query, because these fields are also calculated.
- AutoNumber fields, because the values in AutoNumber fields change only when you add a record to a table.
- Fields in unique-values queries and unique-records queries, because these values in such queries are summarized.
- Fields in a union query, because each record that appears in two or more data sources only appears once in the union query result. Because some duplicate records are removed from the results, Access cannot update all the necessary records.
- Fields that are primary keys, because if they are used in a table relationship, you cannot update the field by using a query unless you first set the relationship to automatically cascade updates.

To create an update query, use the following procedure.

Step 1: Create a select query to select the records that you want to update. This simple example just includes the business phone from the Employees table.

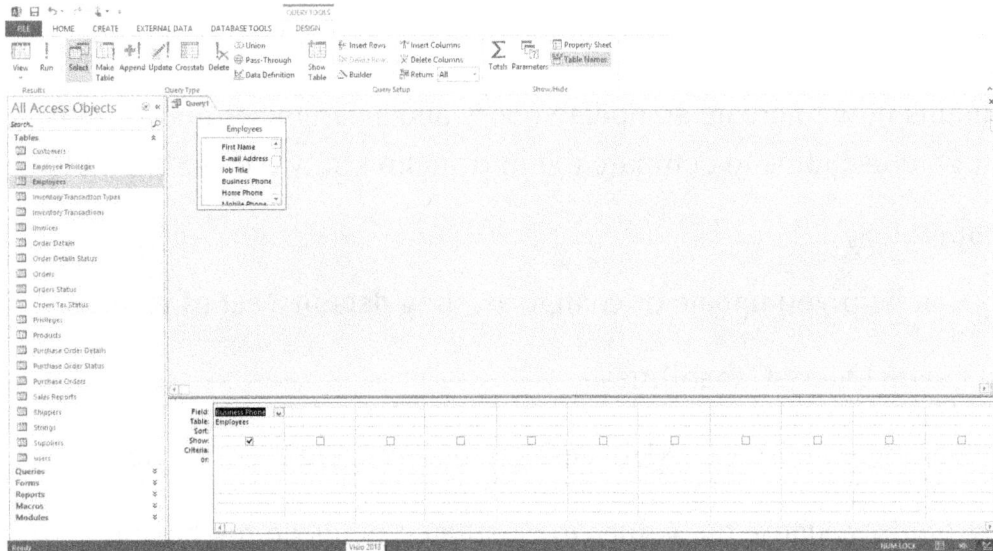

Step 2: Run the query. Make sure that the results include the information that you want to update.

Step 3: Return to the Query Design view.

Step 4: On the Design tab on the Query Tools Ribbon, select Update to convert the query to an Update query.

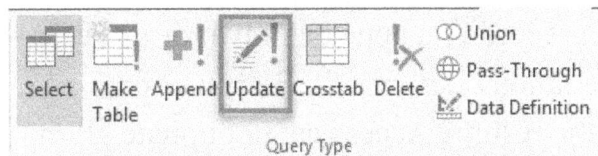

Step 5: Access adds the Update To row in the query design grid. Locate the field that contains the data you want to change.

Step 6: Enter your change criteria. You can use any value expression in the Update To row. The following provide some examples of value expressions:

- "Salesperson" to change a text value to Salesperson
- #8/10/07# to change a date/time field value to 10-Aug-07
- "PN" & [PartNumber] to add "PN" to the beginning of each specified part number
- [UnitPrice] * [Quantity] to multiple the values in fields named UnitPrice and Quantity

150

- If(IsNull([UnitPrice]),0,{UnitPrice}) to change an unknown or undefined value to a 0 in a field named UnitPrice

Field:	Business Phone
Table:	Employees
Update To:	'8008881000'
Criteria:	
or:	

Step 7: Run the query. Access displays a confirmation that you will be updating records. Select Yes to continue.

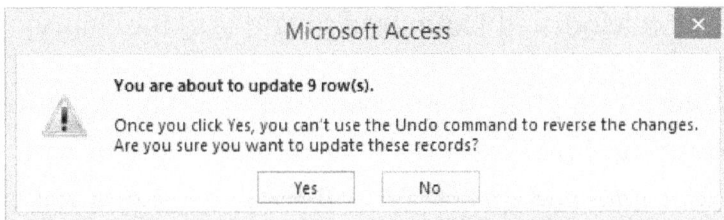

Microsoft Access

You are about to update 9 row(s).

Once you click Yes, you can't use the Undo command to reverse the changes. Are you sure you want to update these records?

Yes   No

Creating a Parameter Query

A parameter query uses user-defined criteria to run the query.

To create a parameter query, use the following procedure.

Step 1: Let's start with the Inventory Purchased query in the sample database. You can find it in the Navigation pane under Supporting Objects. Open it in Design View.

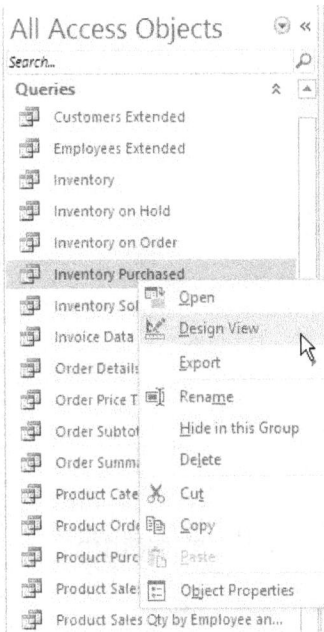

Step 2: Double click on the Transaction Modified Date field to add it to the query.

Step 3: In the Criteria field in the Query Design grid, enter the text that you want to appear on the dialog box for the prompt in brackets. For this example, we will use an expression, which will select a start date and an end date. Here is what you will type in the Criteria field: Between [Start Date] and [End Date]. You can right-click the Criteria field and select Build from the context menu to open the Expression Builder.

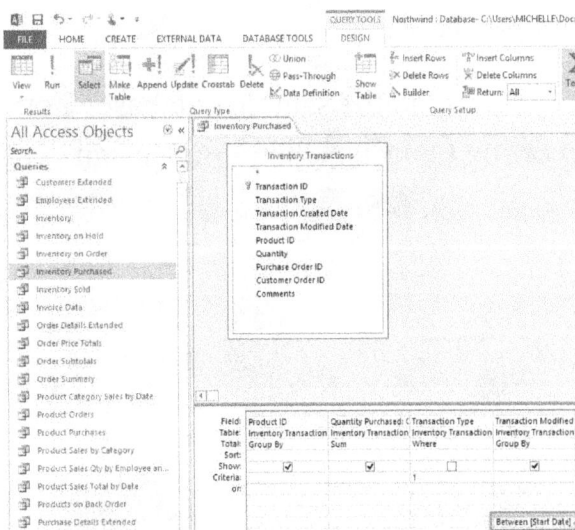

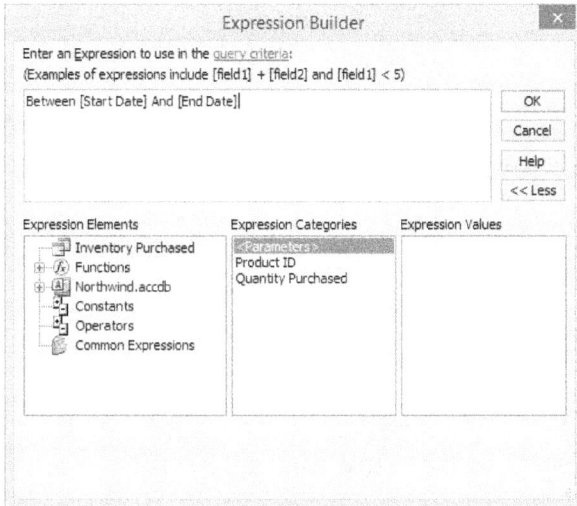

Step 4: Run the query.

Step 5: Access displays the Enter Parameter Value dialog box. Enter the desired criteria. For this example, let's use 04/01/2006 as the start date. Select OK. If you have entered multiple parameters, another dialog box will be displayed. Let's use 4/30/2006 as the end date. Select OK.

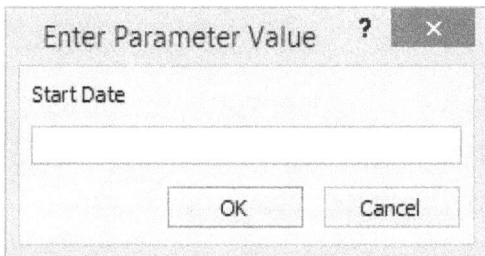

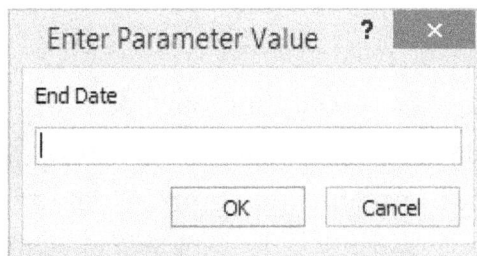

Access displays the results that meet the parameters you entered.

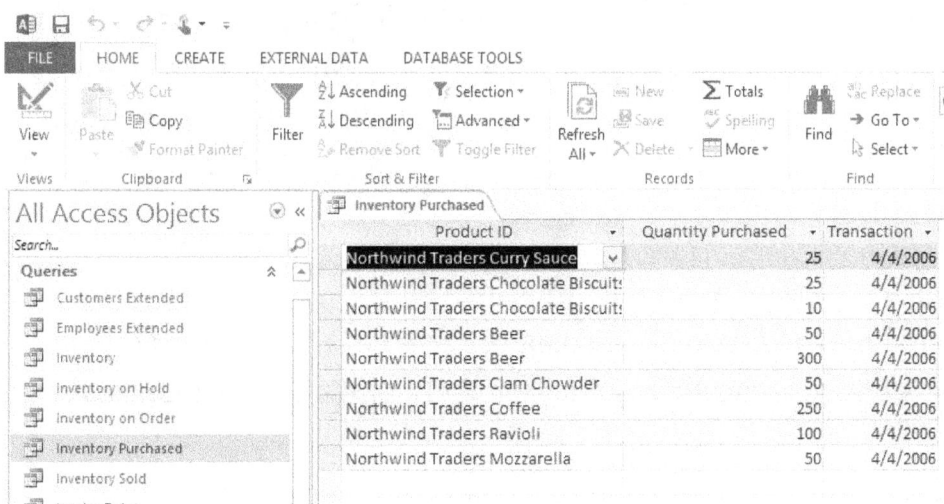

About Joining Data Sources in a Query

You can use a join to compare two tables for matching information.

Joins indicate how data in two sources can be combined based on the data values they have in common. They are very like table relationships. You can join data sources in Query Design view and view or modify the join properties.

If the tables in your query already have a relationship, Access creates an inner join to mirror that relationship. This provides a way to make sure that data pulled from two tables is correctly combined so that the right data from each table is included, without any extraneous data. Access also creates inner joins automatically if you add two tables to a query and both tables each have a field with the same name and the same or compatible data type and if one of the fields is a primary key.

You can create joins to be more inclusive. You should also create joins if you add queries to your query and have not created relationships between those queries.

Types of Joins

In an inner join, Access only includes data from a table if there is corresponding data in the related table, and vice versa. Inner joins are the most common type of join. This type of join lets you combine data from two sources based on shared values.

An outer join is directional because it adds the remaining rows from one of the tables. A left outer join includes all the records from the left table (the first table in the join), and a right outer join includes all the records from the right table. Access does not specifically support a full outer join, which includes all rows from both tables, with rows combined when they correspond. You can achieve this effect with cross joins and criteria.

An unequal join uses an operator other than the equal sign to compare values and determine whether and how to combine the data. Access also does not specifically support unequal joins, but you can achieve the same effect with cross joins and criteria.

A cross join is a side-effect of adding two tables to a query and forgetting to join them. Access interprets this to mean that you want to see every record from one table combined with every record from the other table, or every possible combination of records. This type of join is rarely used. No data can be combined.

To create a join, use the following procedure.

Step 1: Create a new query in Design view.

Step 2: Add the Orders and Order Details tables.

Step 3: Drag the Major field from the Class Enrollments table to the Major field in the Student Majors table.

Step 4: Access joins the two tables on the fields and displays a line between them.

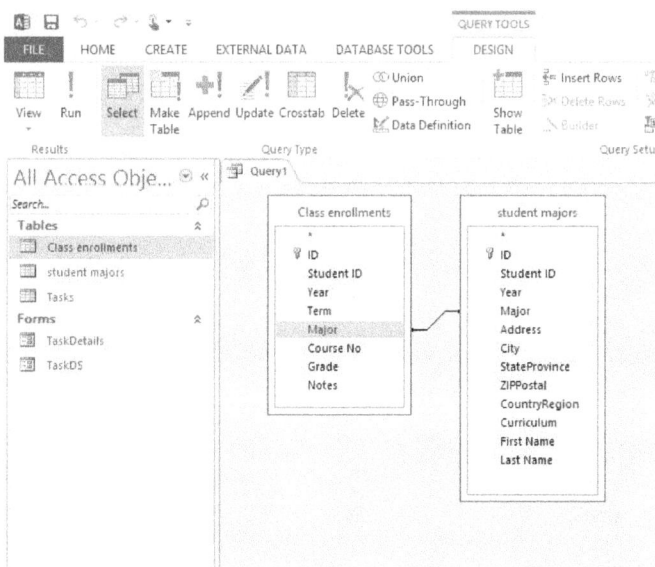

Step 5: Double-click the line to open the Join Properties dialog box.

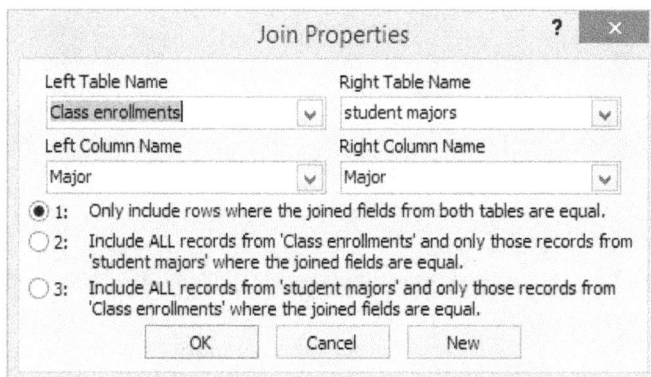

Step 6: The tables and fields are shown in the upper portion of the dialog box.

Step 7: To select an inner join, choose the first option. To select a left outer join, select the second option. The third option creates a right outer join.

Chapter 19: Designing Forms

This chapter will help you become an expert at designing forms. We will start with design view, where you can change the structure of your form. You will also learn how to use form controls, like text boxes, combo boxes, etc. We will look at the Property Sheet, where you can modify anything about your controls. Finally, you will learn how to add header and footer elements to your form, like a logo, title, and the date and time.

Modifying Your Form in Design View

You can change the structure of your form using Design view.

To open a form in design view, use the following procedure.

Step 1: With the form, you want to modify open, select the design icon on the bottom right hand corner of the screen.

You can also open the form from the Navigation pane in design view. Right click on the form to display the context menu. Select Design View.

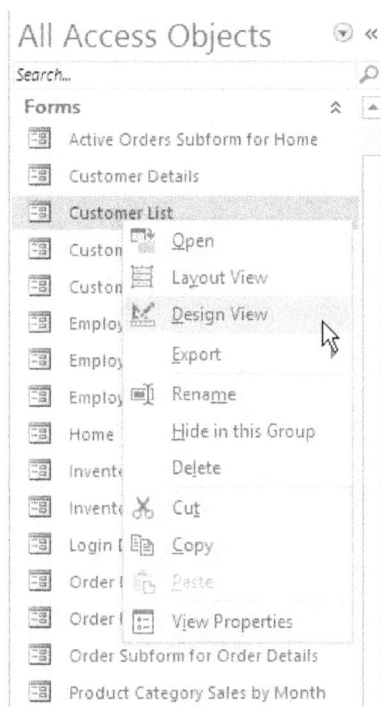

Access opens the form in design view.

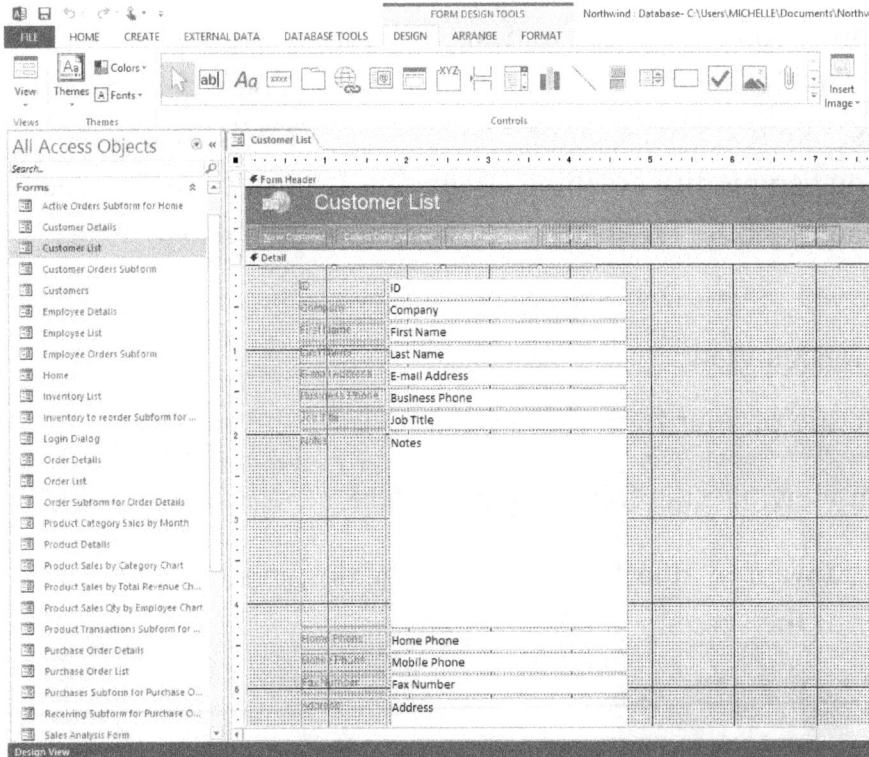

Design view includes the Form Header, the Form Footer, and the Detail sections. (For long forms, you may need to scroll down to see the form footer section.) By default, the grid is displayed to assist with aligning controls.

To change the tab order, use the following procedure.

Step 1: Select the Design tab from the Form Design Tools Ribbon.

Step 2: Select Tab Order.

Access displays the Tab Order dialog box.

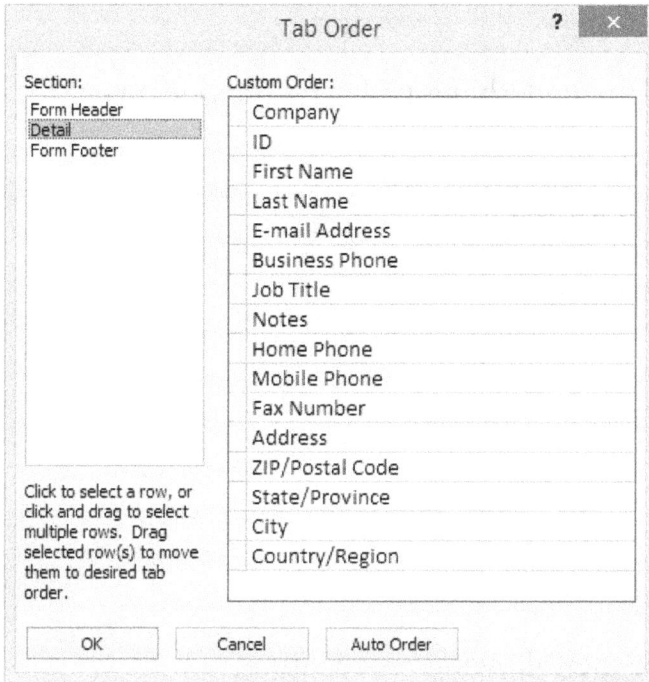

Step 3: To change the order, highlight the row you want to move. To highlight click on the area to the left of the field name.

Step 4: Drag the field to the new location.

Step 5: Select OK when you have finished.

Working with Form Controls

The controls on your form are linked to the underlying table and provide a way for users to enter data.

To create a blank form, use the following procedure.

Step 1: Select the Create tab from the Ribbon.

Step 2: Select Blank Form.

To add a control from the Field List, use the following procedure.

Step 1: Double-click the field in the Field List that you want to add (or drag it from the Field List to the position on the form).

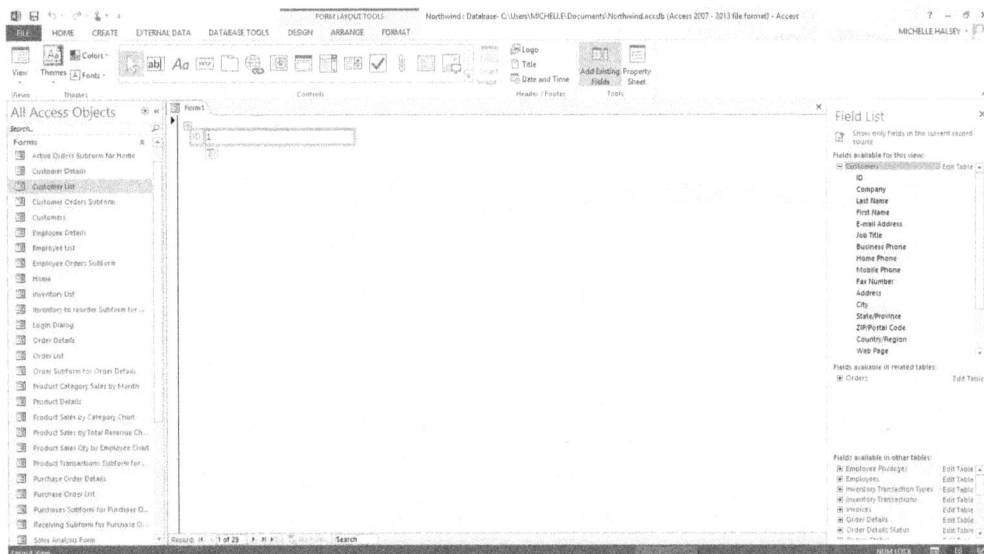

To change the type of control for an item added from the Field List, use the following procedure.

Step 1: Right-click the form control.

160

Step 2: Select Change To from the context menu.

Step 3: Select the type of control you would like to use.

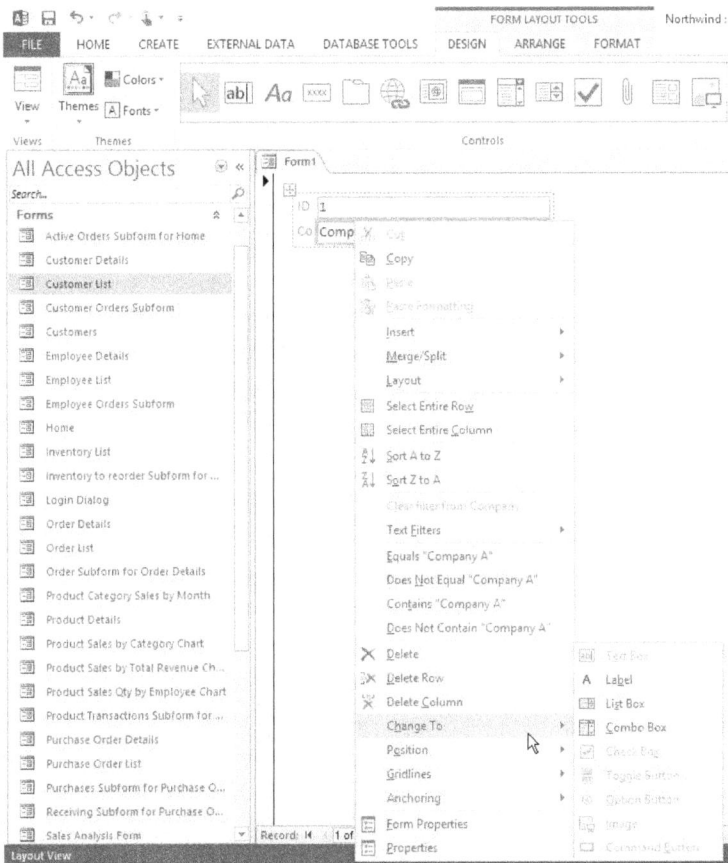

Using the Property Sheet

The Property Sheet allows you to change all the properties for a control. Select Property Street on the Format Layout Tools Design tab.

Here is the Format tab of the Property Sheet for a text box. The Format tab controls how the selected field appears on the form.

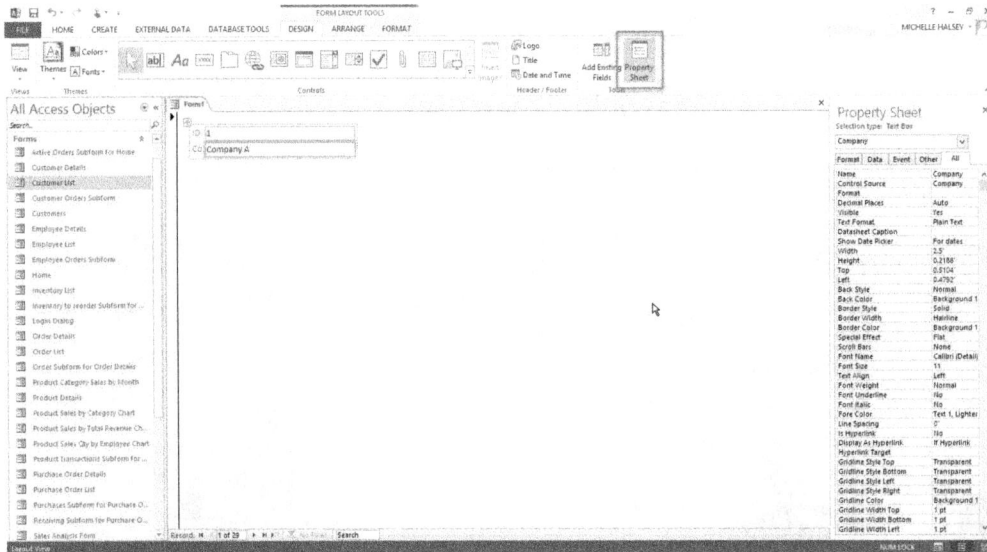

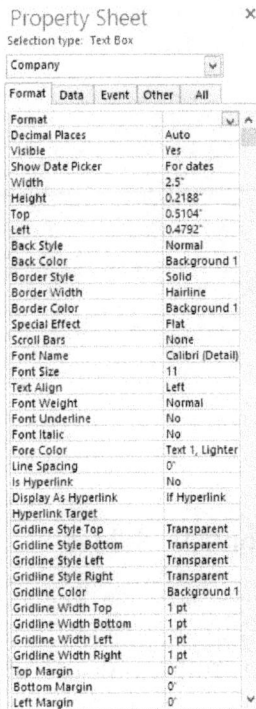

Property Sheet

Selection type: Text Box

Company

Format	Data	Event	Other	All

Format	
Decimal Places	Auto
Visible	Yes
Show Date Picker	For dates
Width	2.5"
Height	0.2188"
Top	0.5104"
Left	0.4792"
Back Style	Normal
Back Color	Background 1
Border Style	Solid
Border Width	Hairline
Border Color	Background 1
Special Effect	Flat
Scroll Bars	None
Font Name	Calibri (Detail)
Font Size	11
Text Align	Left
Font Weight	Normal
Font Underline	No
Font Italic	No
Fore Color	Text 1, Lighter
Line Spacing	0"
Is Hyperlink	No
Display As Hyperlink	If Hyperlink
Hyperlink Target	
Gridline Style Top	Transparent
Gridline Style Bottom	Transparent
Gridline Style Left	Transparent
Gridline Style Right	Transparent
Gridline Color	Background 1
Gridline Width Top	1 pt
Gridline Width Bottom	1 pt
Gridline Width Left	1 pt
Gridline Width Right	1 pt
Top Margin	0"
Bottom Margin	0"
Left Margin	0"

Here is the Data tab of the Property Sheet for a text box. The Data tab controls where the data comes from for the selected field on the form. Here is where you can set a validation rule and message text.

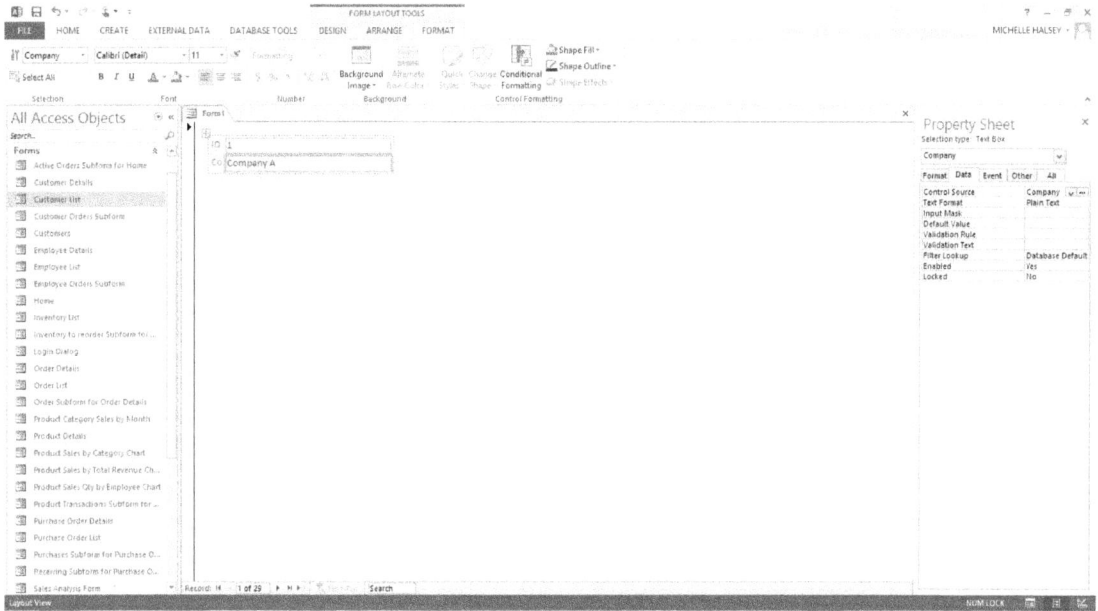

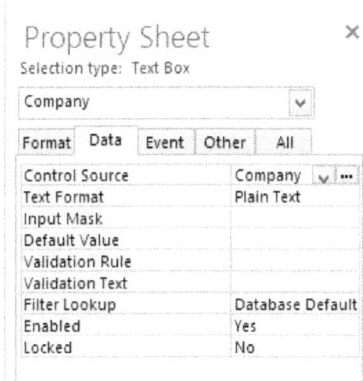

Here is the Event tab of the Property Sheet for a text box. The Event tab allows you to program macros connected to the control, such as something that happens when you click on the field on the form.

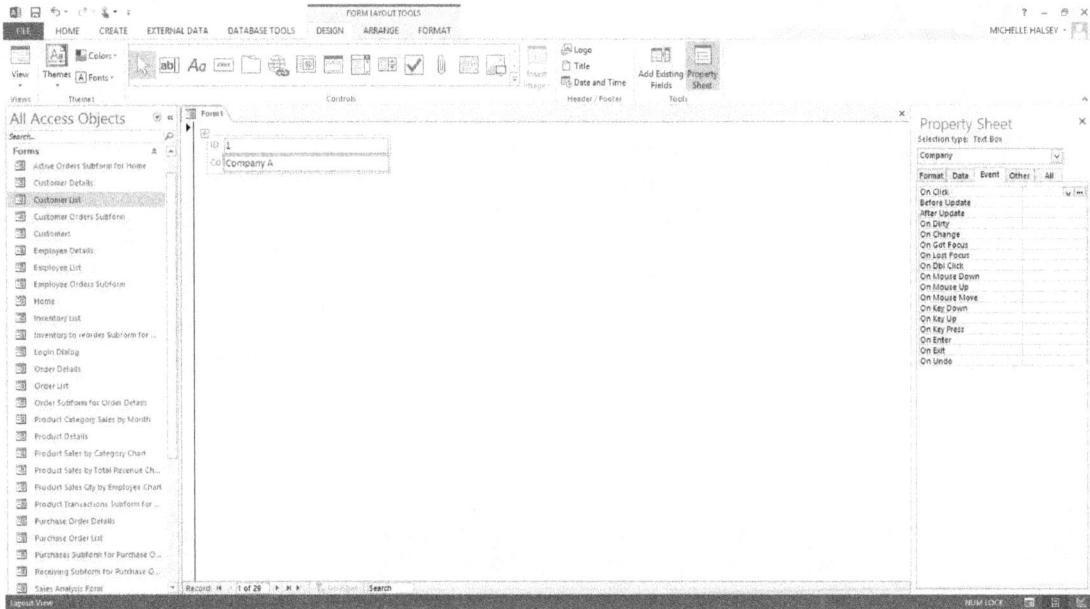

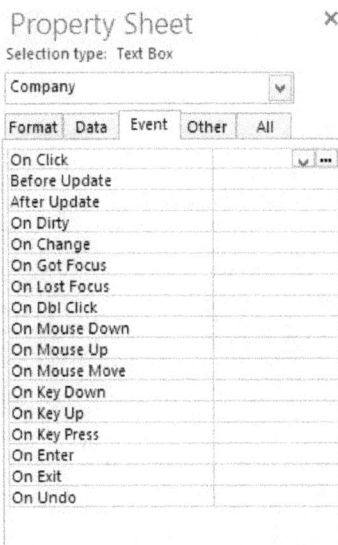

Here is the Other tab of the Property Sheet for a text box. The Other tab includes additional information used for the selected field on the report.

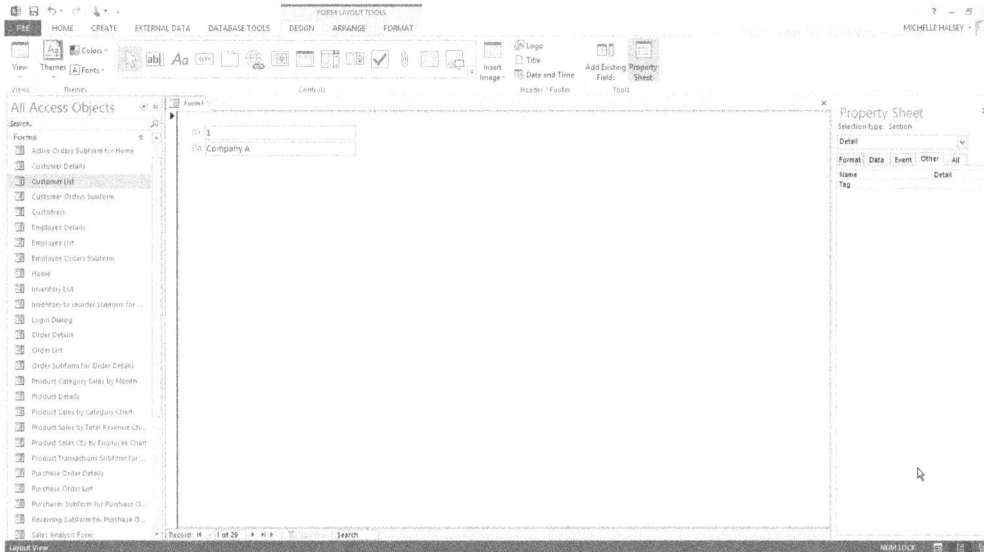

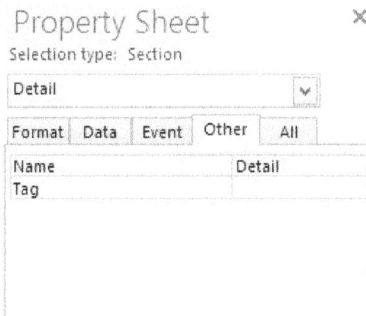

The All tab of the Property Sheet includes all the previous information on one tab, in case you want to program everything about a field at once.

Adding Header and Footer Elements

You can add a logo, a title, and the date and time to your form as header or footer elements.

To add a header and footer to the form, use the following procedure.

Step 1: Click the Design ribbon of the Form Layout Tools Ribbon.

Step 2: Select Logo in the Header/Footer menu to add a Logo to the header, select Logo.

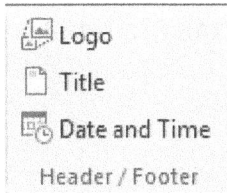

Step 3: Navigate to the location of the picture file that you want to use. Highlight and select OK in the Insert Picture dialog box.

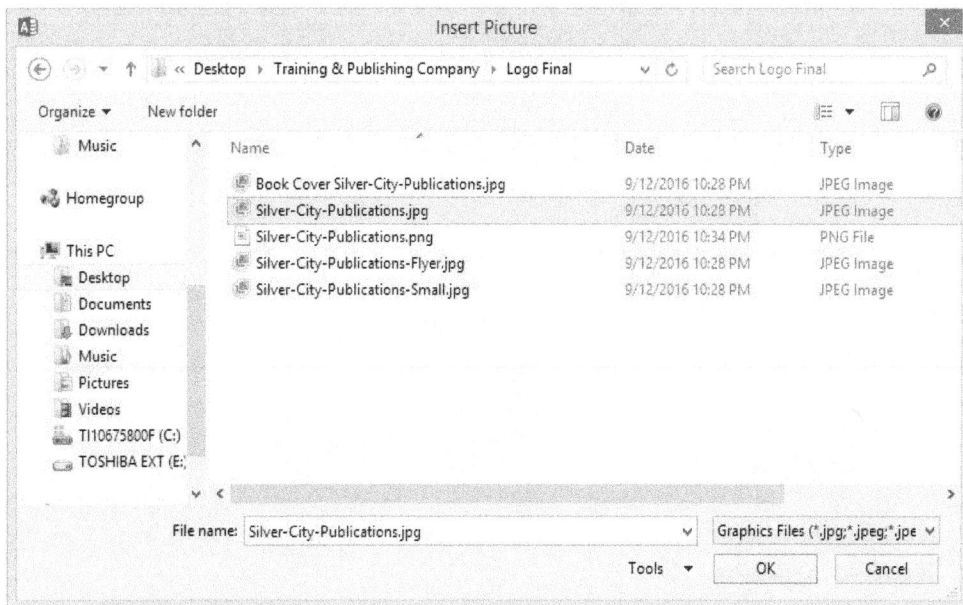

Step 4: The picture is added in the Form Header.

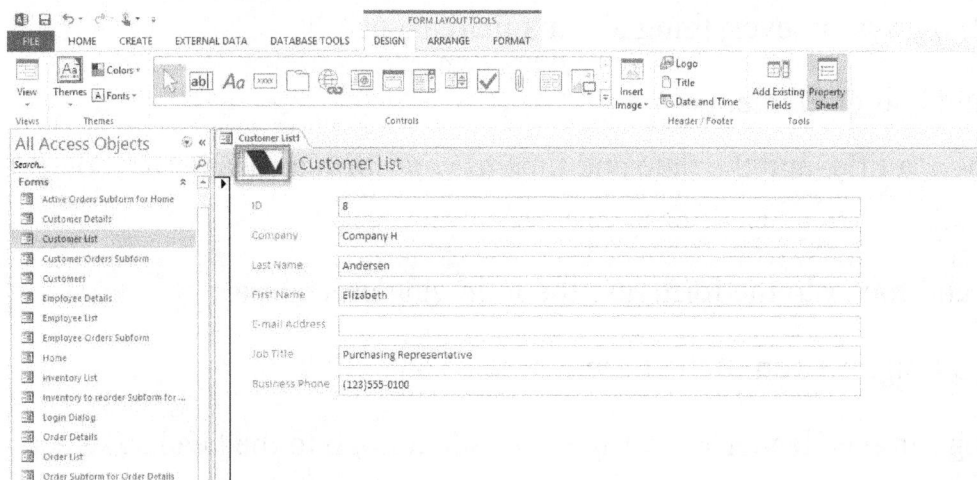

Step 5: To add a title to the form, select Title.

166

Step 6: Enter the Title for the form.

Customer List1	
Customer List	
ID	8
Company	Company H
Last Name	Andersen
First Name	Elizabeth
E-mail Address	
Job Title	Purchasing Representative
Business Phone	(123)555-0100

Step 7: To add the date and time, select Date and Time.

Step 8: In the Date and Time dialog box, you can choose whether you want to include the date and/or the time and which format you want to use for each element. Select OK.

Date and Time   ?   ×

☑ Include Date
◉ Monday, August 7, 2017
○ 07-Aug-17
○ 8/7/2017

☑ Include Time
◉ 10:03:41 PM
○ 10:03 PM
○ 22:03

Sample:
Monday, August 7, 2017
10:03:41 PM

[OK]   [Cancel]

Step 9: You can drag the header elements to the footer, if desired.

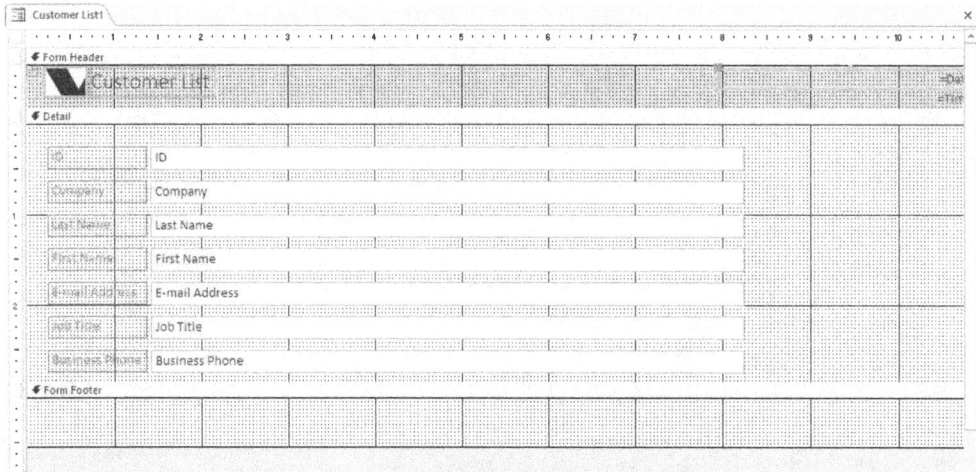

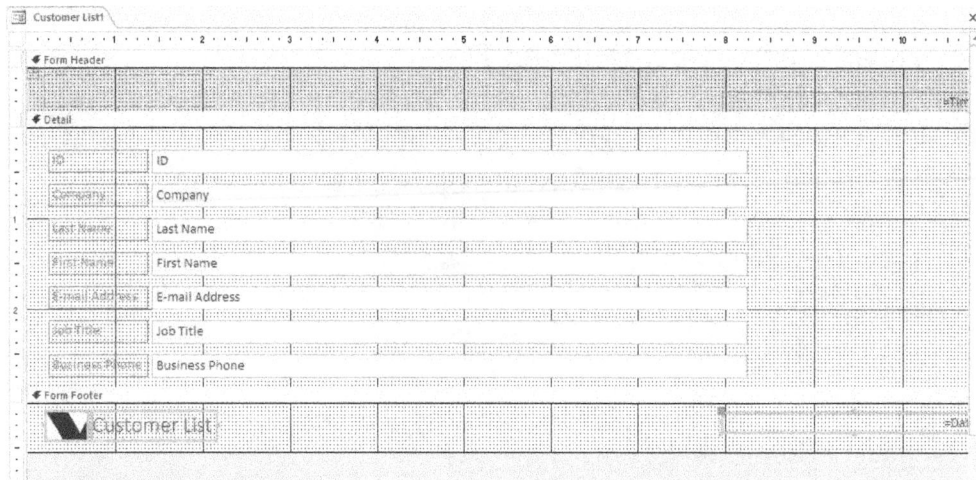

You can preview your form design by selecting Form View from the View menu on the Design tab of the Form Layout Tools Ribbon.

Chapter 20: Advanced Reporting Tasks

This chapter will help you become an expert at designing reports. Again, we will start with the design view for reports. Then you will learn how to apply conditional formatting. You will also learn how to group and sort items in a report. This chapter discusses how to add calculated controls. Finally, we will look at creating labels.

Using Report Design View

Just as with the other objects, Design View for reports gives you the ability to control several structural items for your report.

To review the design view for the Invoice report, use the following procedure.

Step 1: Right click the Invoice report in the Navigation pane and select Design view from the context menu.

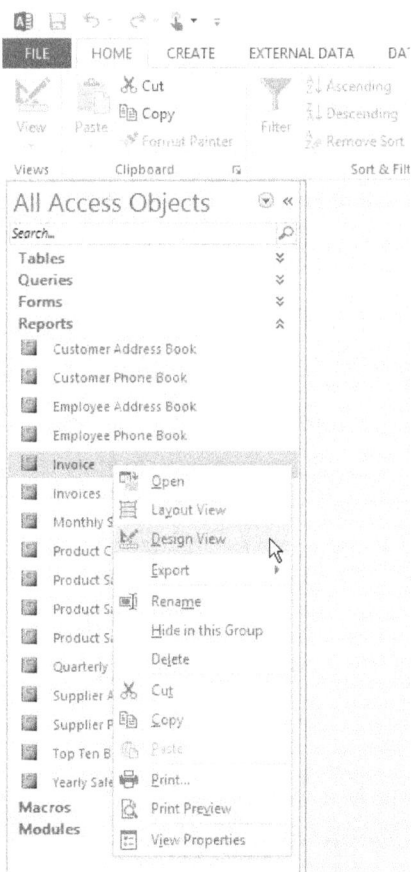

This is what you will see. Notice the report sections we talked about. This report is grouped by Order ID, so that is the title of the Group Header and Group Footer.

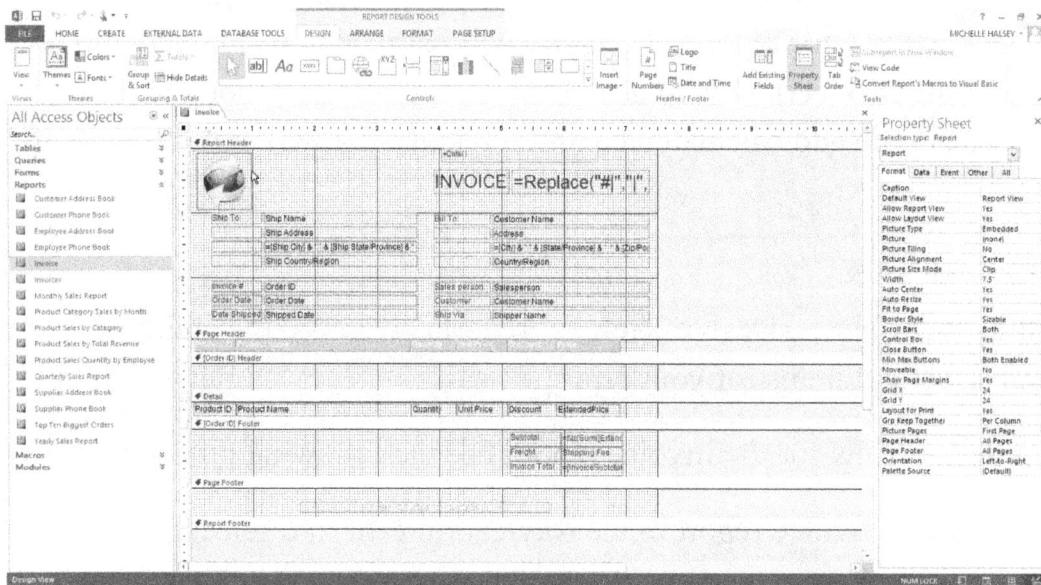

Use the report header when you want to insert items on in the header of your report cover page. Items that you might want to include on a cover page header include the title of the report or a related logo or image.

Use the page header for items that you want to insert into the header of every single page of your report. For example, you may want to have the title of the report on each page.

Use the group header to signal the beginning of a new group. The group header will be placed at the beginning on each group. A good example of something to put as the group header is the name of the group.

Use the details section for the controls that make up the body of your report.

Use the group footer to signal the end of a group. The group footer will be placed at the end of each group. You may want to use this space to summarize the details of the group.

Use the page footer for items that you want to show at the bottom of each page. A common use of page footers is to display page numbers or the section/group name that the page is found under.

Use the report footer to summarize the details of the entire report or to report a total if your report pertains to expenses, profit/loss, etc.

Using Conditional Formatting

Conditional formatting allows you to automatically format certain items.

To create conditional formatting rules, use the following procedure.

Step 1: Select the Format tab from the Report Design Tools Ribbon.

Step 2: Select Conditional Formatting.

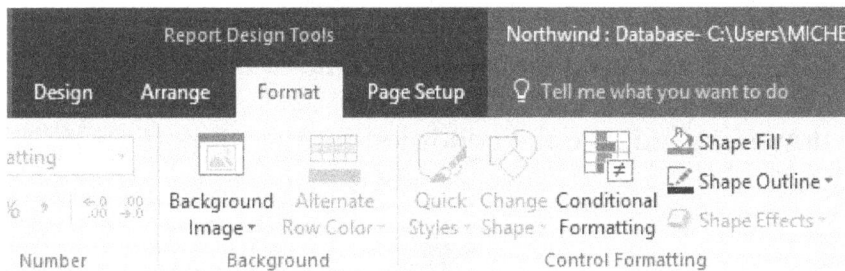

Step 3: In the Conditional Formatting Rules Manager dialog box, select New Rule to create a new rule.

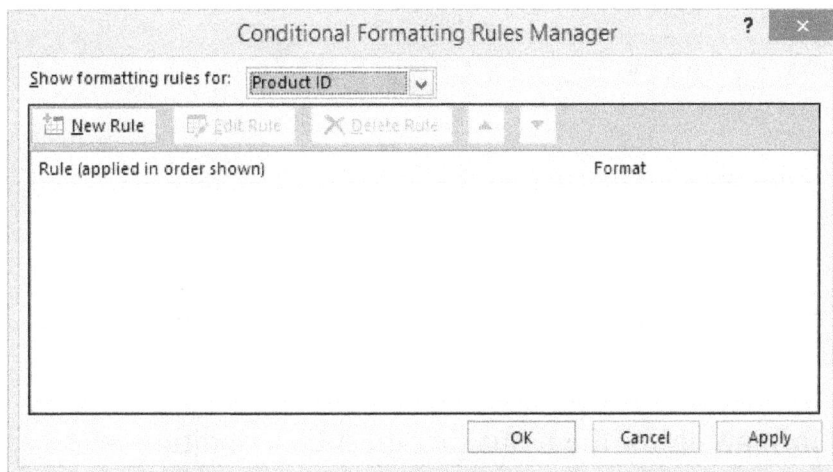

Step 4: In the New Formatting Rule dialog box, select the Rule Type.

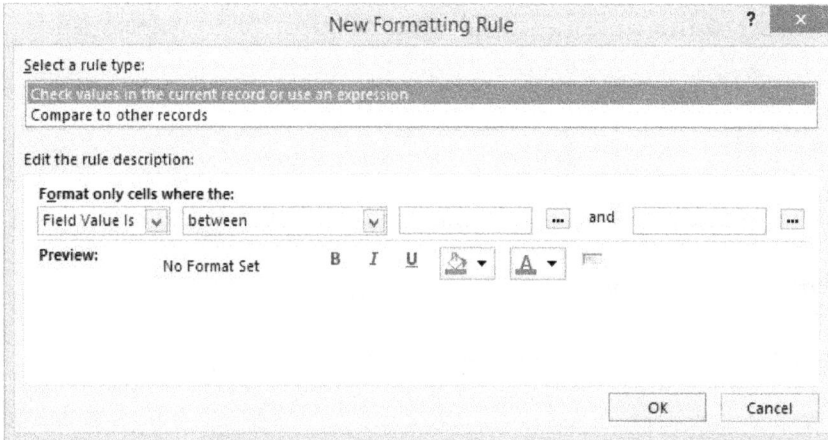

Step 5: Select an option from the first Format only cells ... drop down list to indicate whether the formatting will be based on the field value or an expression.

Step 6: In the second drop down list, indicate the condition for the formatting based on the first selection.

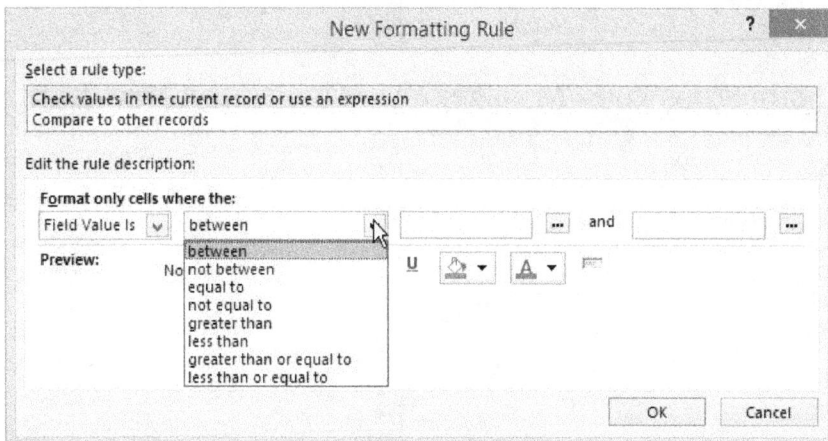

Step 7: Complete the rule by placing the values or expressions you want to use in the two blank fields.

Step 8: Finally, change the formatting for the items that meet the condition in the rule using the Formatting tools. The Preview area shows an example of how the items will be formatted.

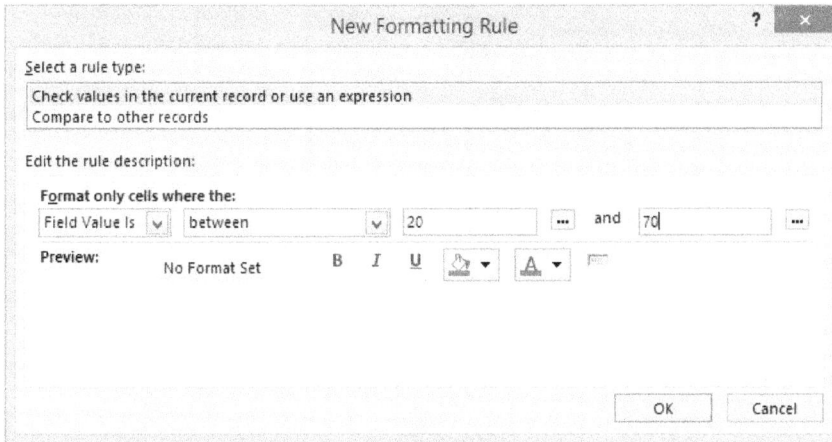

New Formatting Rule

Select a rule type:

Check values in the current record or use an expression
Compare to other records

Edit the rule description:

Format only cells where the:

| Field Value Is | between | 20 | ... | and | 70 | ... |

Preview: No Format Set B I U

OK   Cancel

Step 9: Select OK to close the New Formatting Rule dialog box.

Step 10: You can create additional rules, or select OK to close the Conditional Formatting Rules Manager.

Using the Group, Sort, and Total Pane

The Group, Sort and Total pane can help you add or modify grouping, sorting, and totals for a report.

To open the Group, Sort, and Total pane, use the following procedure.

Step 1: Select the Design tab from the Report Layout Tools Ribbon.

Step 2: Select Group & Sort.

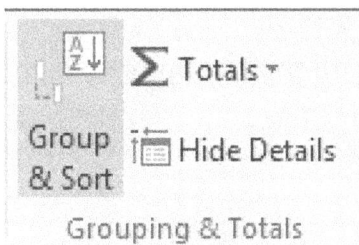

The Group, Sort, and Total pane appears at the bottom of the screen. Notice that this example already has a grouping.

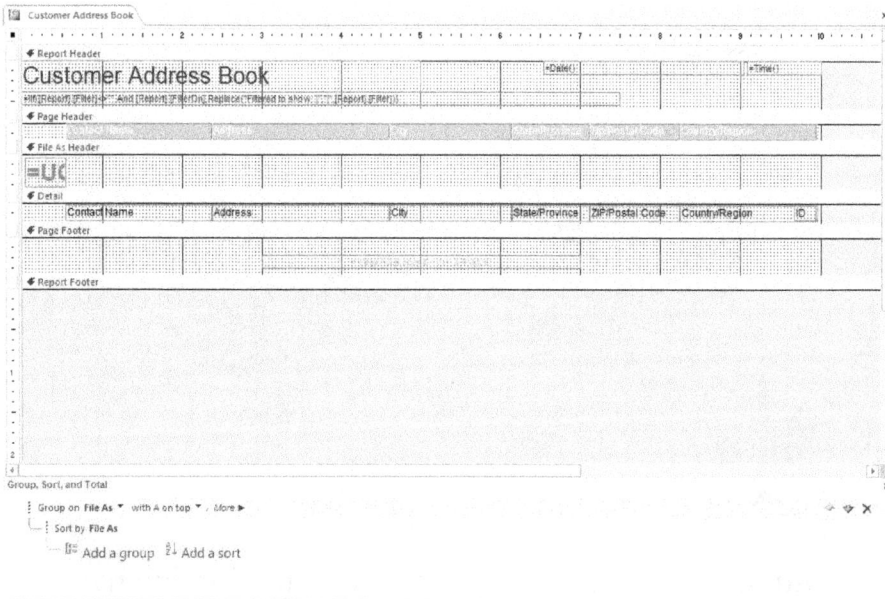

Click on More to see the details for this grouping.

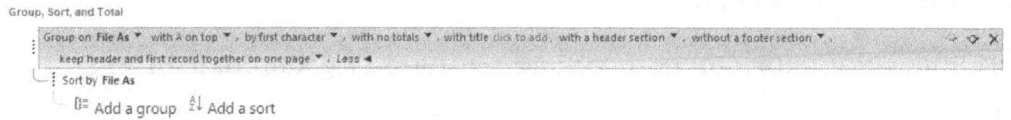

Let's add an additional grouping. Select Add a group.

Another level appears on the Group, Sort, and Total pane with the Group on select field and the list of fields open. Let's group the customers by state.

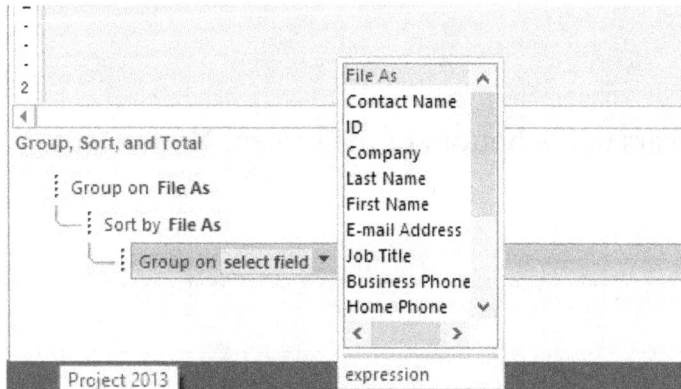

Notice if you click More, there are additional options, each with a drop-down list of selections, that can help you customize the grouping or sorting.

This is not quite how we want it. Let's get the state grouping first. Use the arrows on the right of the pane to move the group information up or down.

Now we have an alphabetical list of customers by state. You can further customize the grouping by formatting the labels and controls, just as with any other report items.

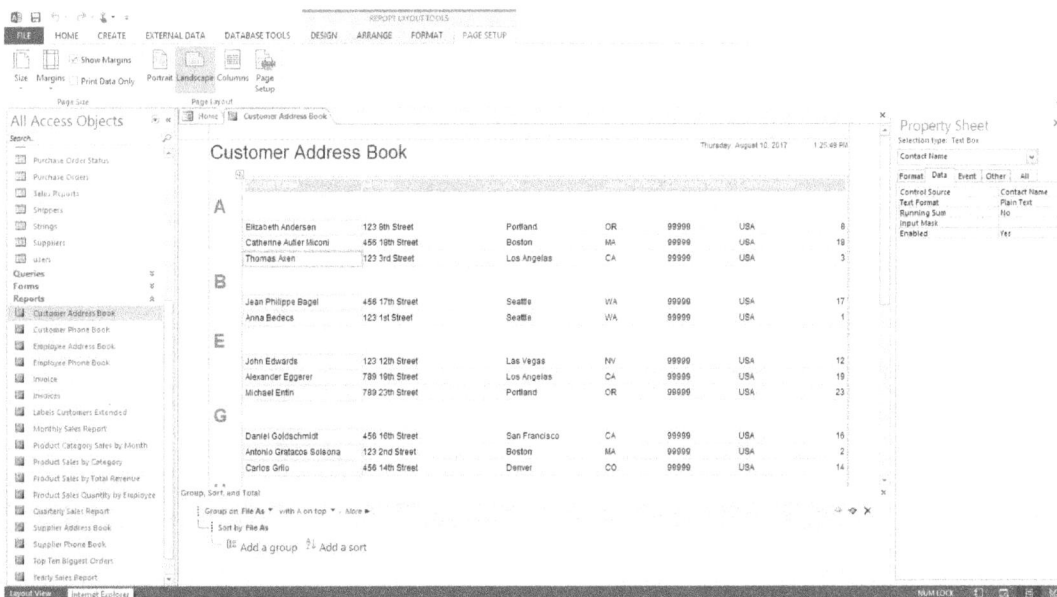

Adding Calculated Controls

Reports can show calculated controls.

To add a calculated control, use the following procedure. In this example, we will add a total to the Top Ten Biggest Orders report and investigate the expression used for the total.

Step 1: Open the Top Ten Biggest Orders report in Layout View.

Step 2: Using the Group, Sort, and Total pane, add a Sum and a Total for the SaleAmount.

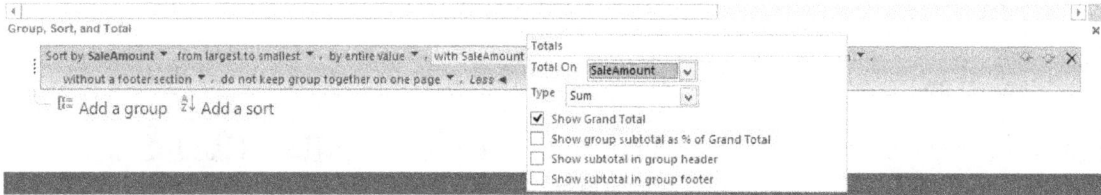

Step 3: Now click on the control that Access created for the total.

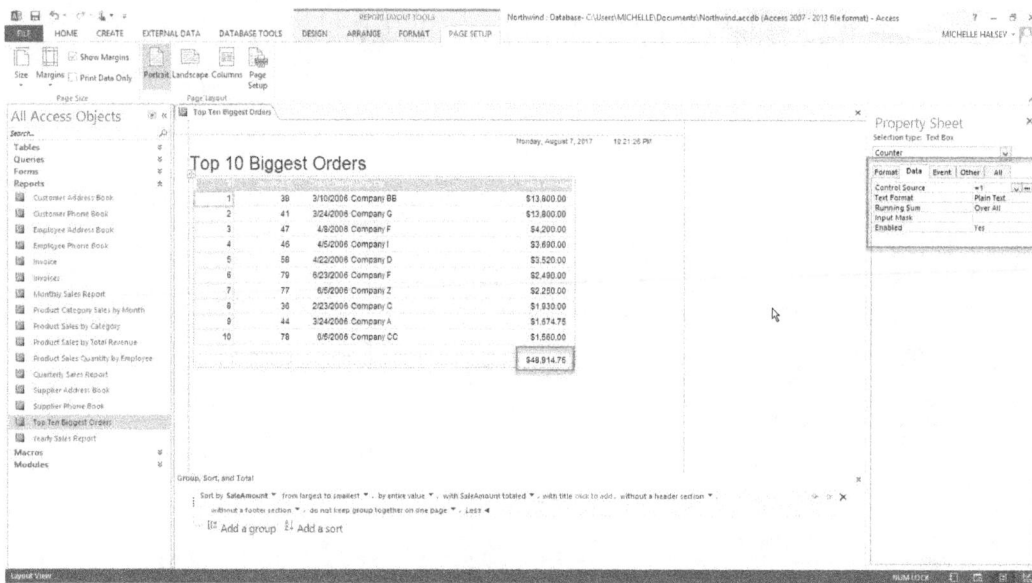

Step 4: If the Property Sheet is not already open, select Property Sheet from the Report Layout Tools Design tab.

Step 5: Select the Data tab.

Step 6: Review the Control Source.

Step 7: If you wanted to create your own calculated control, you could click the … button to open the Expression Builder to create your calculation.

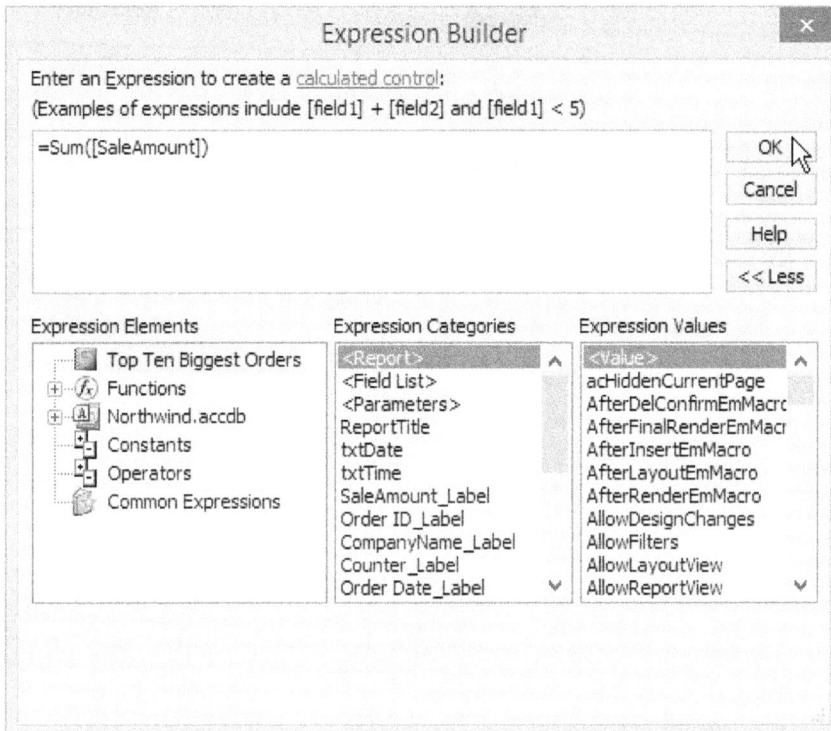

Creating Labels

The Labels option in Access allows you to print mailing labels.

To create mailing labels, use the following procedure.

Step 1: Select the Customer Address Book in the Navigation pane to show which data you want to use for the labels.

Step 2: Select the Create tab from the Ribbon.

Step 3: Select Labels.

Step 4: In the Label Wizard dialog box, select the Product number for the type of labels you would like to use when printing your labels. If you do not find your manufacturer, you can choose Customize to enter the label size. Select Next.

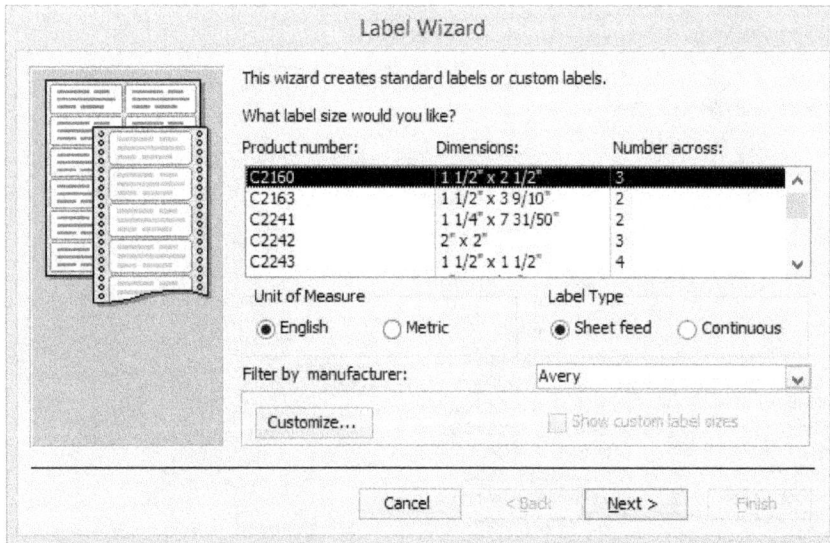

Step 5: In the next screen, select the Font Name, Font Size, Font Weight, and Text Color from the drop-down lists. You can check the Italic and/or Underline boxes if desired. Select Next.

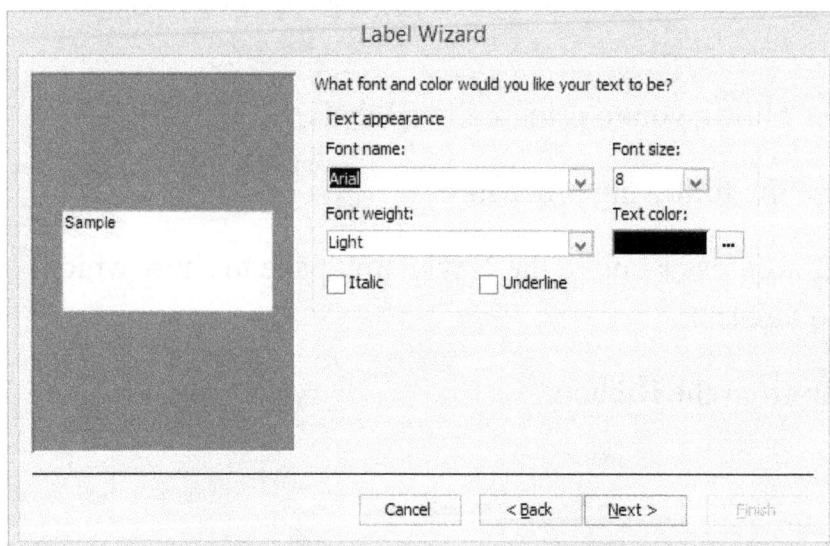

Step 6: The third screen allows you to determine which data you want to print on the labels. Double-click the field from the left list to add it to the right. Press Enter to go to a new line. Select Next.

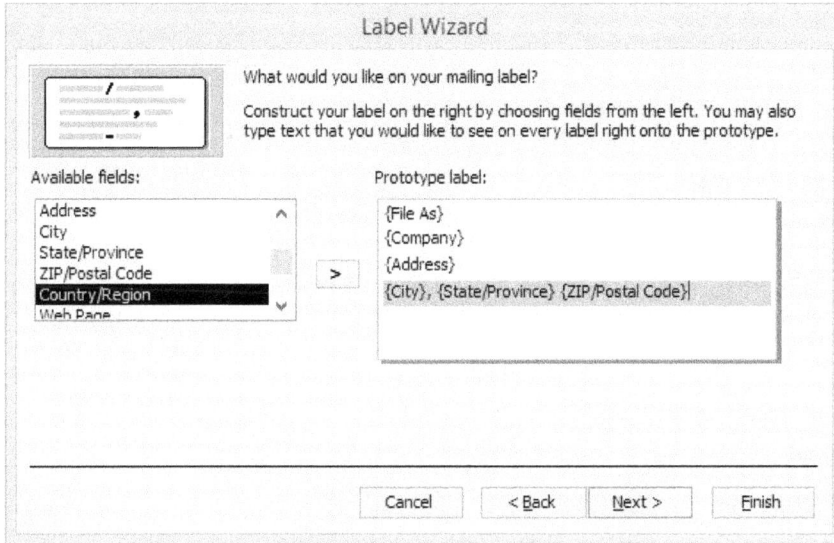

Step 7: If you would like a sort order, select one or more fields from the left list and add them to the right list. Select Next.

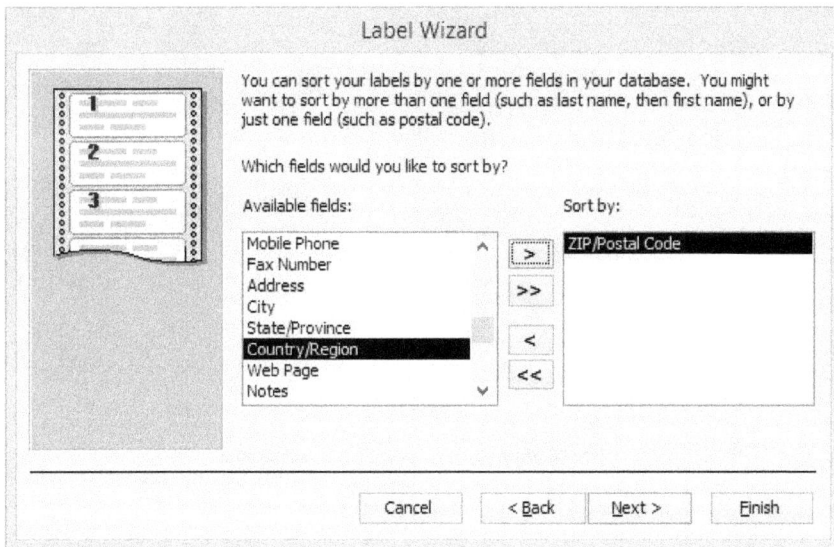

Step 8: In the final screen, enter a name for the label report. Select whether to preview or print the labels or modify the label design. Select Finish.

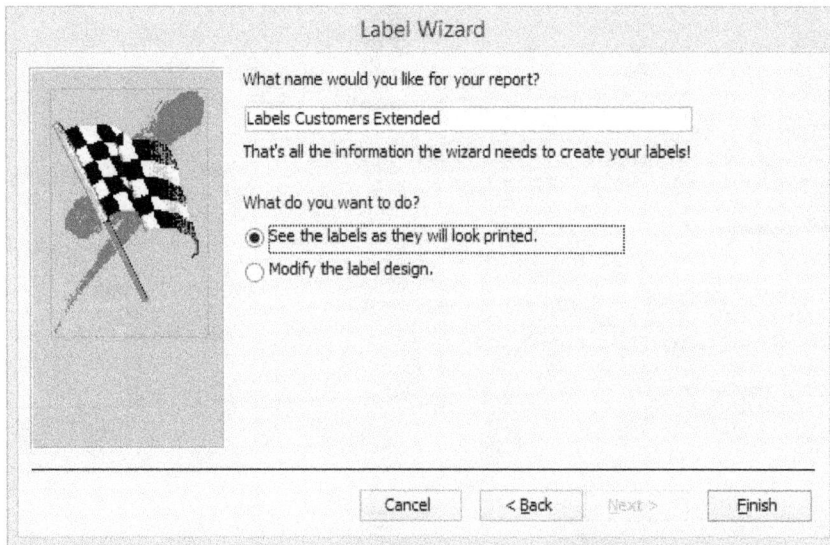

Note that you can modify the design in Report Design view.

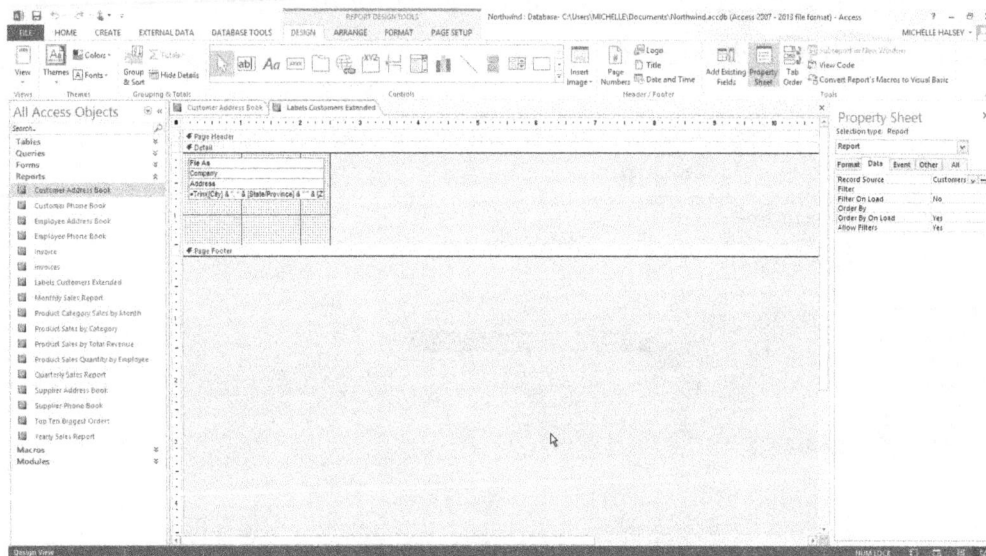

Chapter 21: Advanced Database Tools

In this chapter, we will look at several tools to help you improve your database. First, we will look at the database documenter, which can help you keep a record of your database objects for offline reference. Next, we will look at two ways to analyze your database. The Table Analyzer helps with tables that contain too many repetitions of the same data that should be separated for better performance. The Database Performance tool provides specific suggestions for improving your database. Last, we will look at the compact and repair tool.

Using the Database Documenter

The Database Documenter will help you to document your database for reference and planning.

To use the database documenter, use the following procedure.

Step 1: Select the Database Tools tab from the Ribbon.

Step 2: Select Database Documenter.

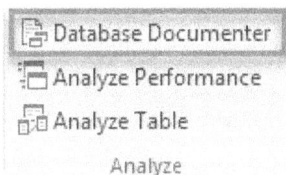

Step 3: In the Documenter dialog box, there are tabs for Tables, Queries, Forms, Reports, Macros, Modules, Current Database, and All Object Types. Select the tab that has the database object that you want to document, or select All Object Types for a list of everything. Check the boxes next to each part of the database that you want to print, or choose the Select All option to select all objects in the current tab.

Tables Tab

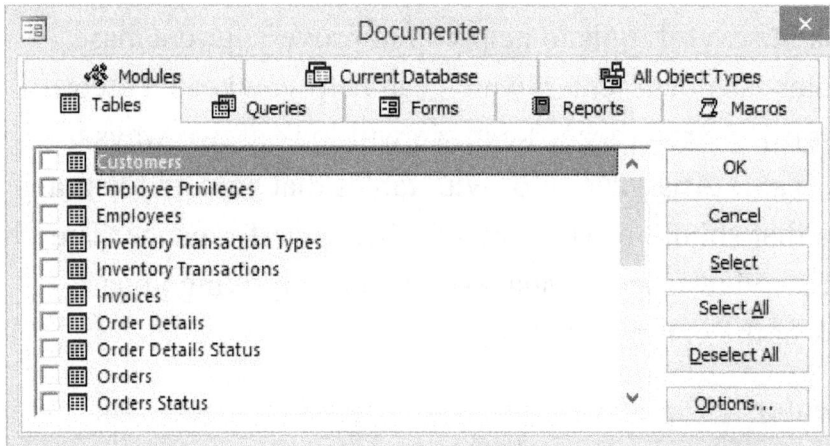

Queries Tab

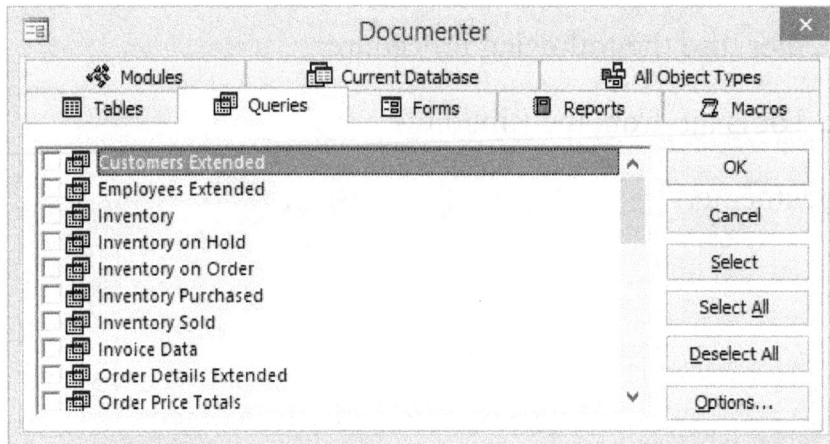

Form Tab

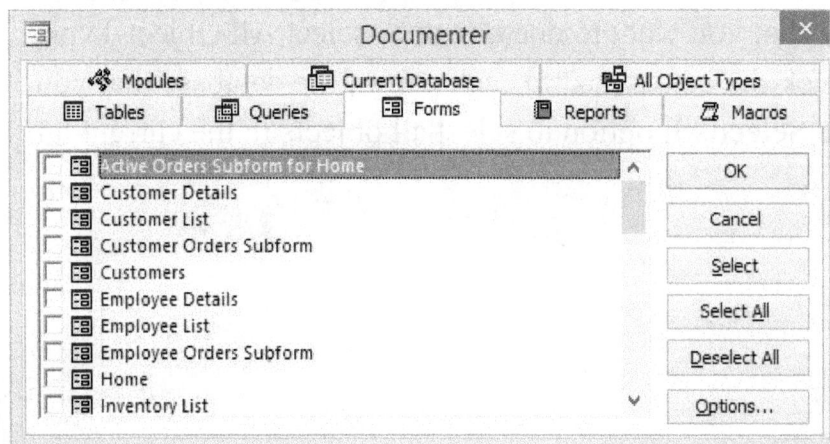

Reports Tab

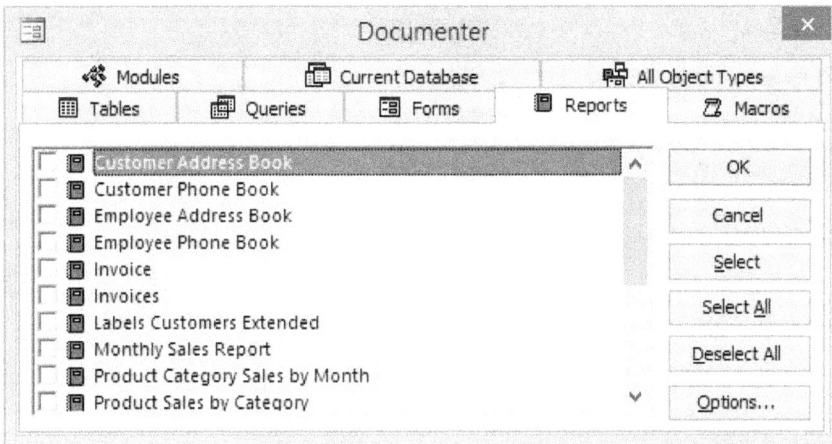

Macros Tab

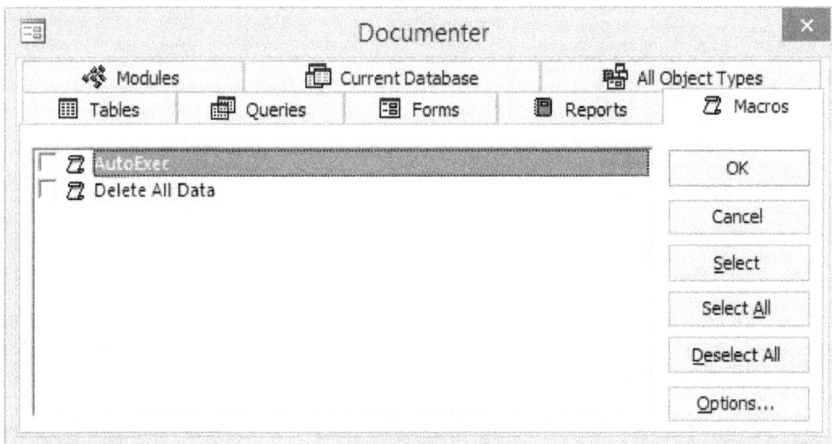

Modules Tab

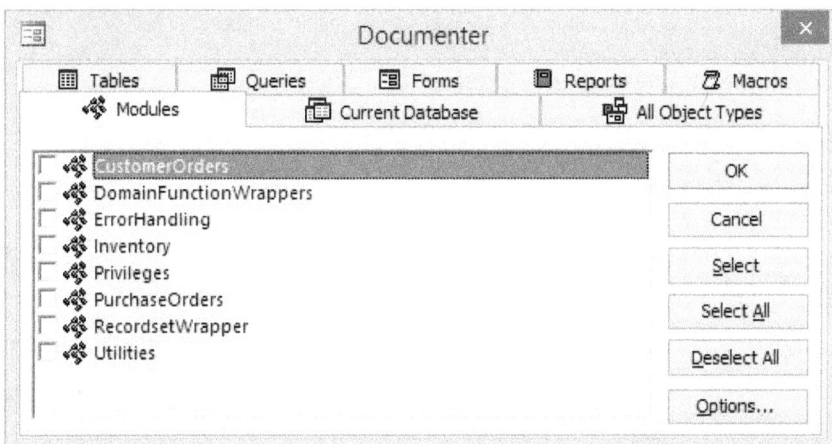

Current Database Tab

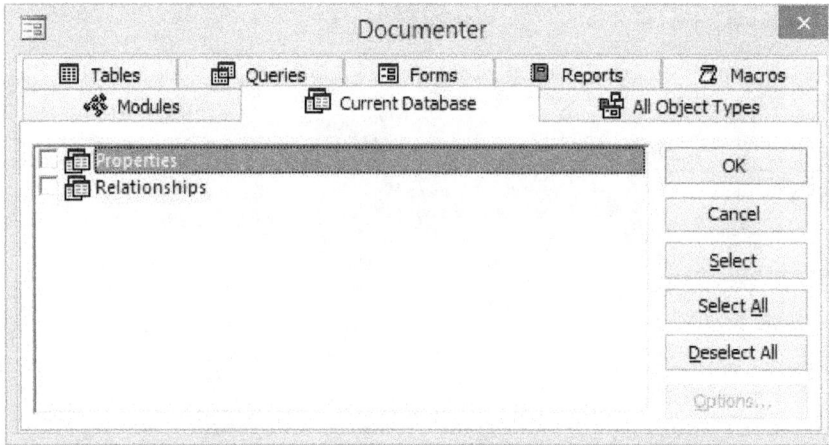

All Object Types

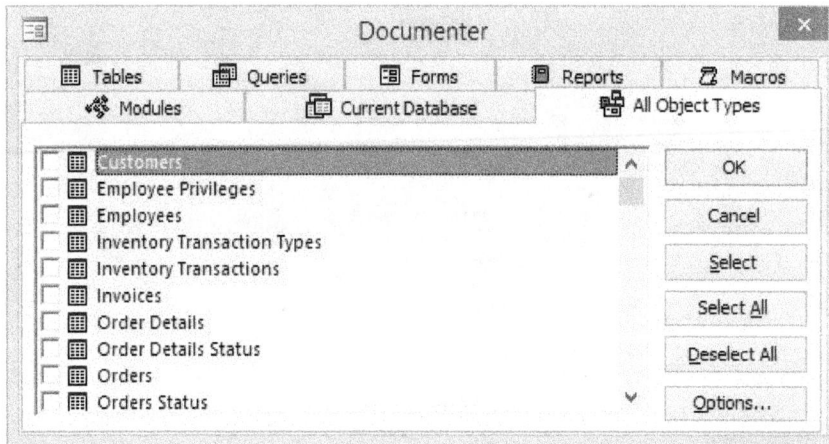

Step 4: Select OK.

The Database Documenter creates a report containing the detailed data for the objects you selected. The report opens in Print Preview.

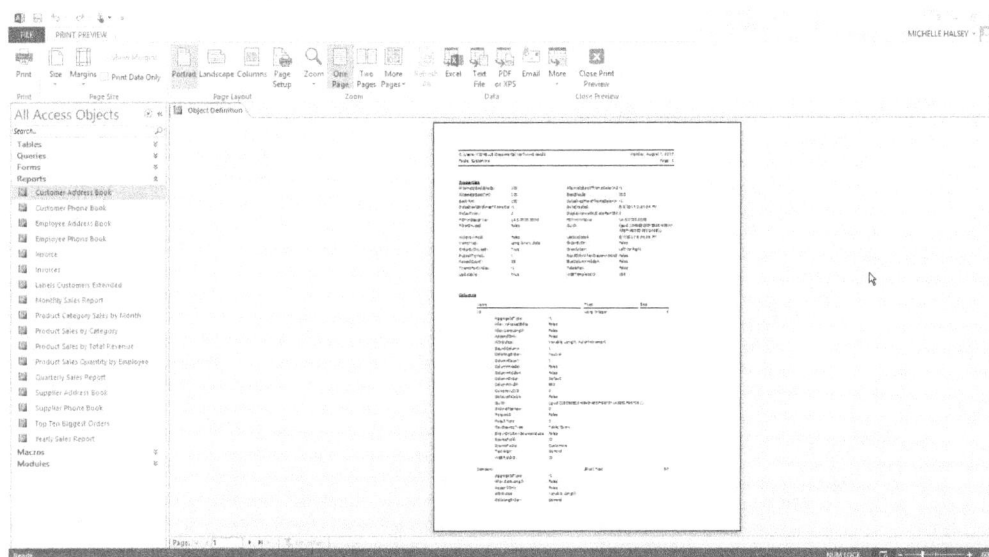

You can view one, two or more pages at a time. You can also change the size and margins, insert columns or switch to landscape formatting before you print. Zoom into the view by clicking on the preview. Select Print to print the documentation.

Analyzing Table Performance

The Table Analyzer helps you to normalize your imported data into properly split out related tables.

To analyze performance, use the following procedure.

Step 1: Select the Database Tools tab from the Ribbon.

Step 2: Select Analyze Table.

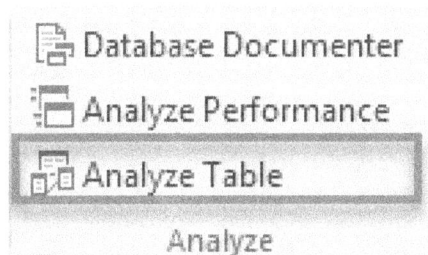

Step 3: The first two screens of the Table Analyzer Wizard are introductory. Select Next.

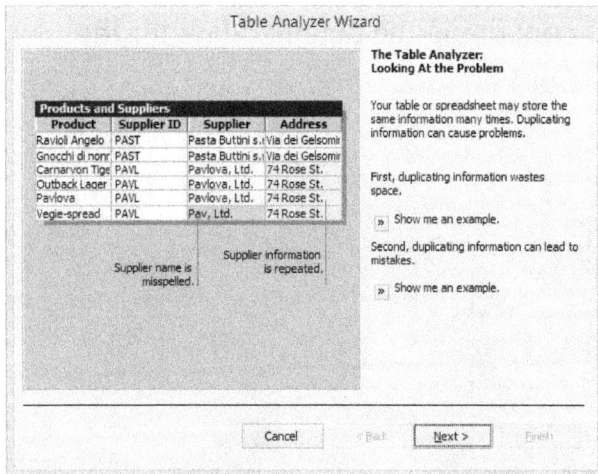

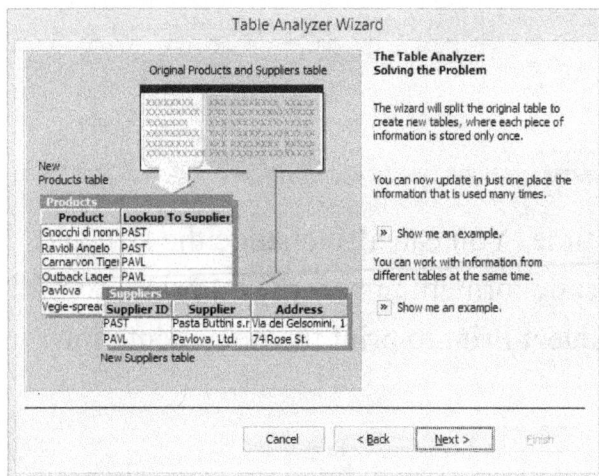

Step 4: On the next screen, select the option for deciding what fields go in what table. Select Next.

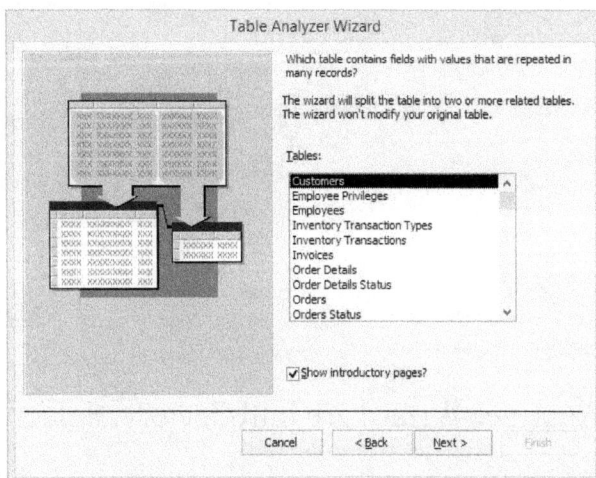

Step 5: Select the option for the wizard to decide which fields to include in the table.

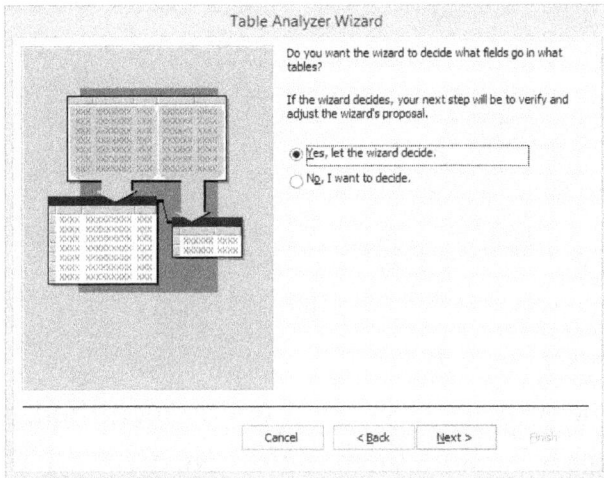

Step 6: The next screen shows the tables the Analyzer will create, with the fields, relationships, etc. You can click the small icon on the top right to rename the selected table. Select OK to name the table. Select Next.

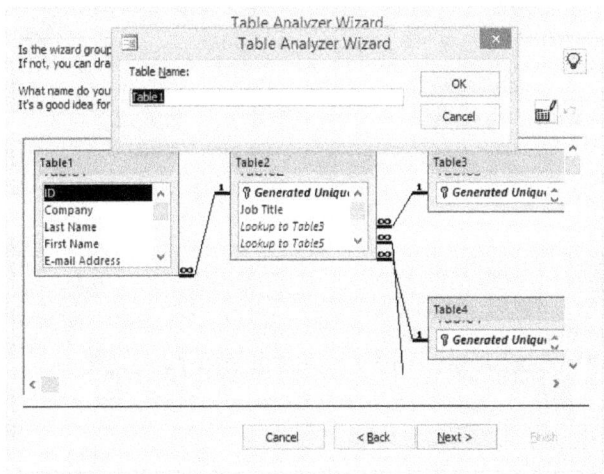

Step 7: The next screen controls the grouping for the split information. You can drag and drop fields to form groups that make sense. Select Next.

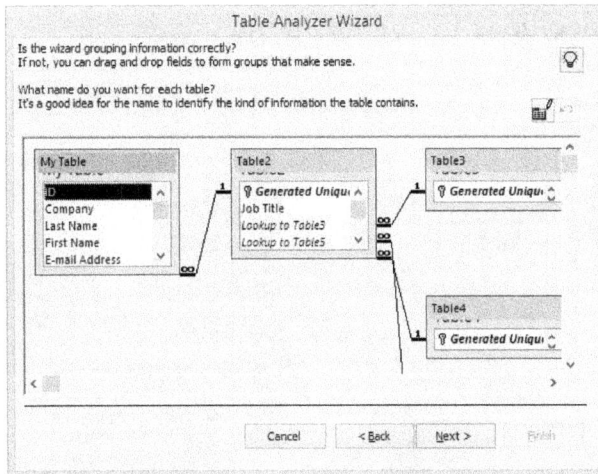

Step 8: In the next screen, make sure that the primary keys are properly identified. You can select the small icon at the top right to rename them. Select Next.

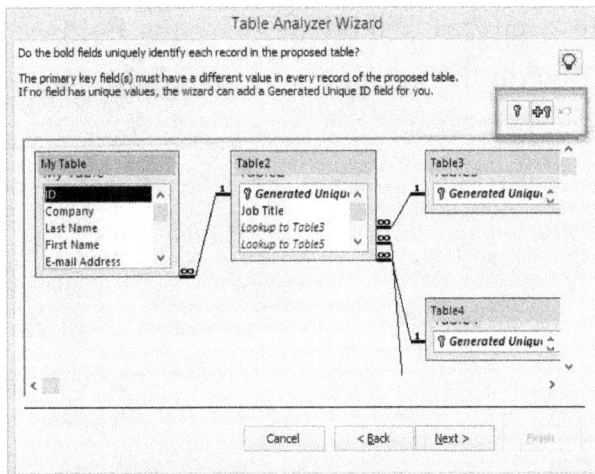

Step 9: On the last screen, indicate whether Access should create a query. Select Finish.

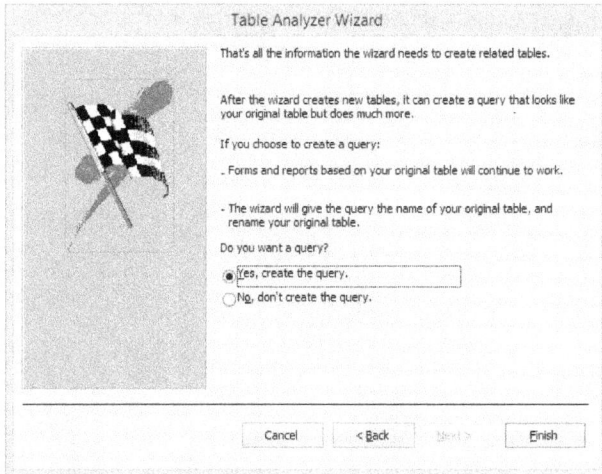

Analyzing Database Performance

The Performance Analyzer opens a window like the Database Documenter dialog box to select objects to analyze.

To analyze database performance, use the following procedure.

Step 1: Select the Database Tools tab from the Ribbon.

Step 2: Select Analyze Performance.

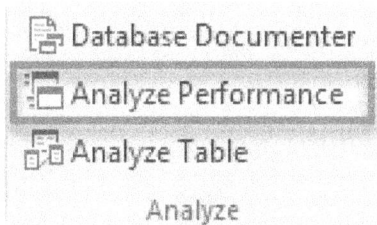

Step 3: In the Performance Analyzer dialog box, there are tabs for Tables, Queries, Forms, Reports, Macros, Modules, Current Database, and All Object Types. Select the tab that has the database object that you want to analyze, or select All Object Types for a list of everything. Check the boxes next to each part of the database that you want to analyze, or choose the Select All option to select all objects in the

current tab. In this example, we will select the Current Database tab and the Relationships box.

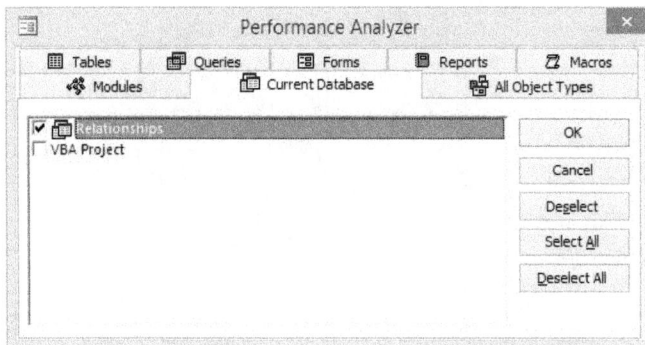

Step 4: The Analysis Results are displayed. There are 4 result categories:

- Recommendation
- Suggestion
- Idea
- Fixed

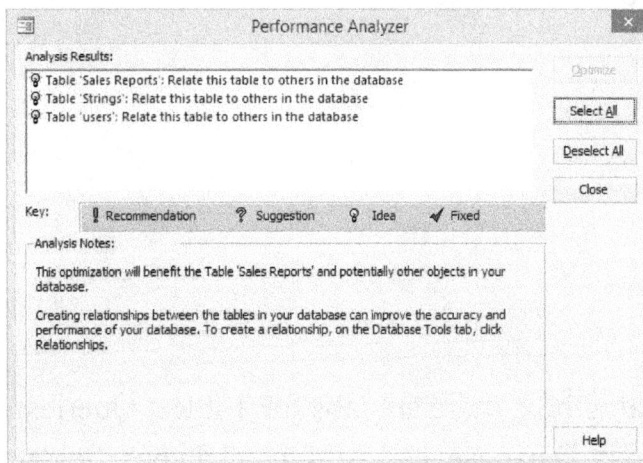

Step 5: Select the recommendation, suggestion or idea that you want to implement. Select Optimize.

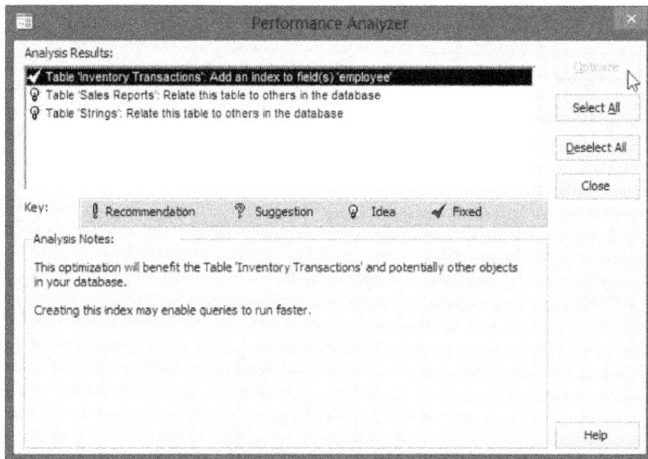

Access makes the change and shows the Fixed icon next to the selected item.

Compacting and Repairing a Database

Database files shared over a network may be subject to occasional corruption. The Compact and Repair command can help.

To compact and repair a database, use the following procedure.

Step 1: Select the File tab on the Ribbon.

Step 2: Select Compact & Repair Database.

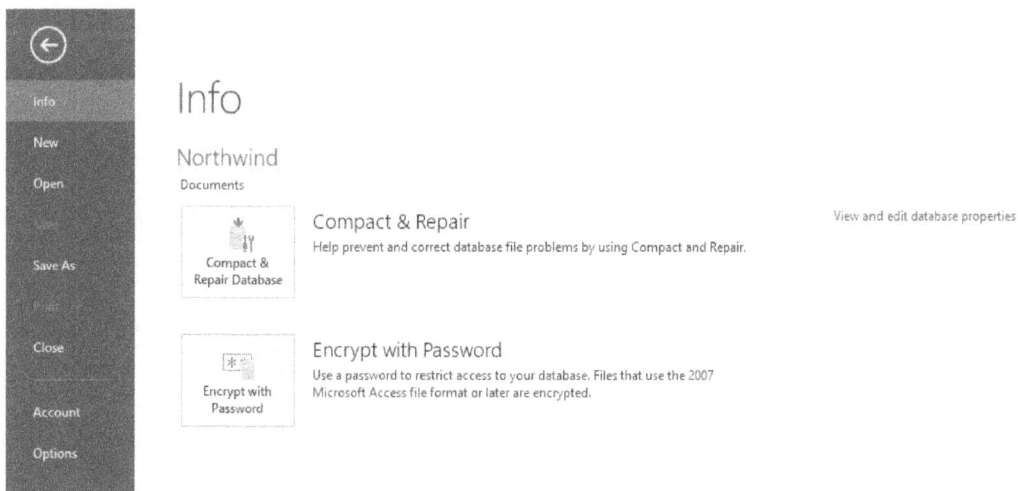

Chapter 22: Working with the Access Environment

This chapter will get you more confident with customizing the Access Environment. First, we will look at the Database Properties dialog box, where you can attach different types of information to a database to help with organizing your files or to help with searching for a file. Then we will look at how to encrypt a database with a password. Then, you will learn how to use the Save Object As command. Finally, we will look at customizing Access by setting options.

Working with Database Properties

Database properties help you set the metadata.

To open the document properties, use the following procedure.

Step 1: Select the File tab from the Ribbon to open the Backstage View.

Step 2: Select View and edit database properties from the right side of the screen.

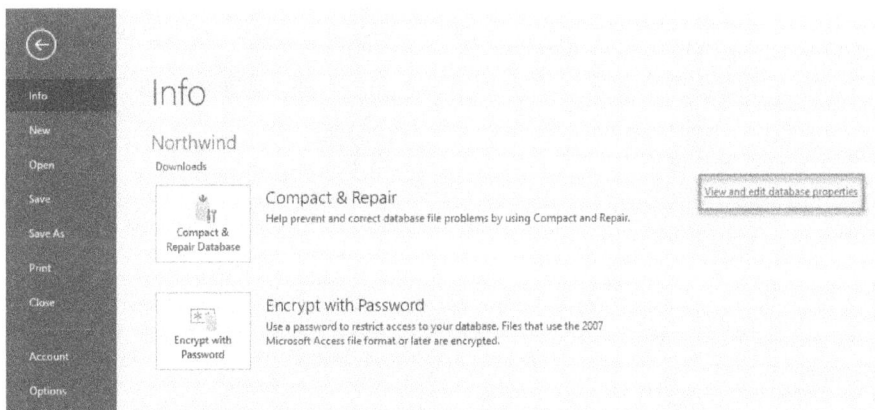

The Database Properties dialog box opens to the Summary tab, where you can add the following details about the database:

- Title
- Subject
- Author (you can type over the default)
- Manager
- Company

- Category
- Keywords
- Comments
- Hyperlink base

The General, Statistics, and Contents tabs allow you to view additional information about the database.

General Tab

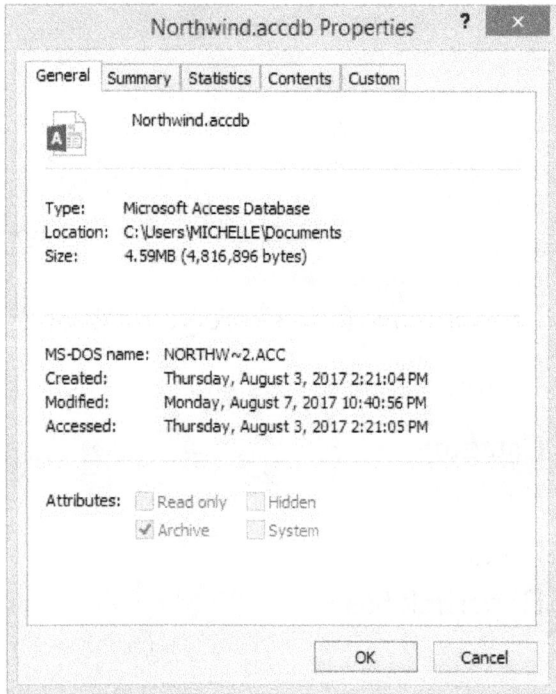

Statistic Tab

Northwind.accdb Properties   ? ✕

| General | Summary | Statistics | Contents | Custom |

Created: Thursday, August 3, 2017 2:20:18 PM
Modified: Monday, August 7, 2017 10:40:56 PM
Accessed: Thursday, August 3, 2017 2:21:05 PM
Printed:

Last saved by:
Revision number:
Total editing time:

OK   Cancel

Contents Tab

Northwind.accdb Properties   ? ✕

| General | Summary | Statistics | Contents | Custom |

Document
contents:

Tables
    Customers
    Employee Privileges
    Employees
    Inventory Transaction Types
    Inventory Transactions
    Invoices
    Order Details
    Order Details Status
    Orders
    Orders Status
    Orders Tax Status
    Privileges
    Products
    Purchase Order Details
    Purchase Order Status
    Purchase Orders
    Sales Reports
    Shippers
    Strings

OK   Cancel

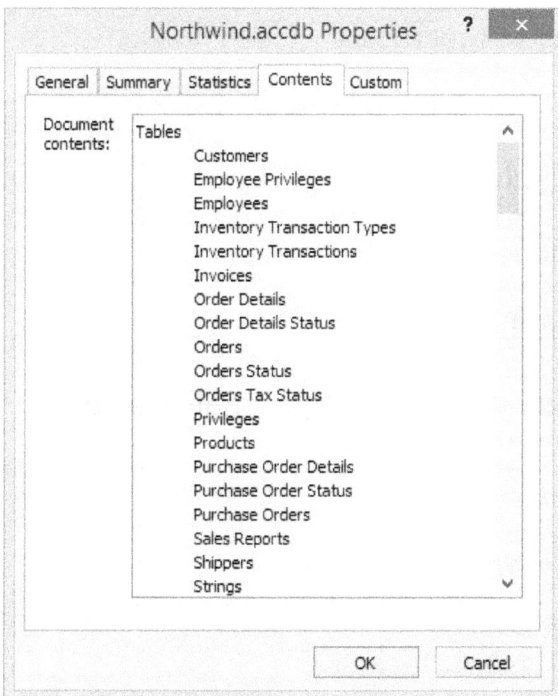

To include customized details about the database, select the Custom tab.

Step 1: Select a Name from the list or type a new name and select Add.

Step 2: Select the type of information to include from the Type drop down list.

Step 3: Enter the information you want to include in the Value field.

Select OK to save any changes to the Document Properties.

Encrypting a Database with a Password

Setting a password encrypts the database for protection.

To open a database with exclusive access, use the following procedure.

Step 1: Use the Open command, not the Recent list.

Step 2: In the Open dialog box, navigate the file you want to open. Select arrow next to Open. Select Open Exclusive.

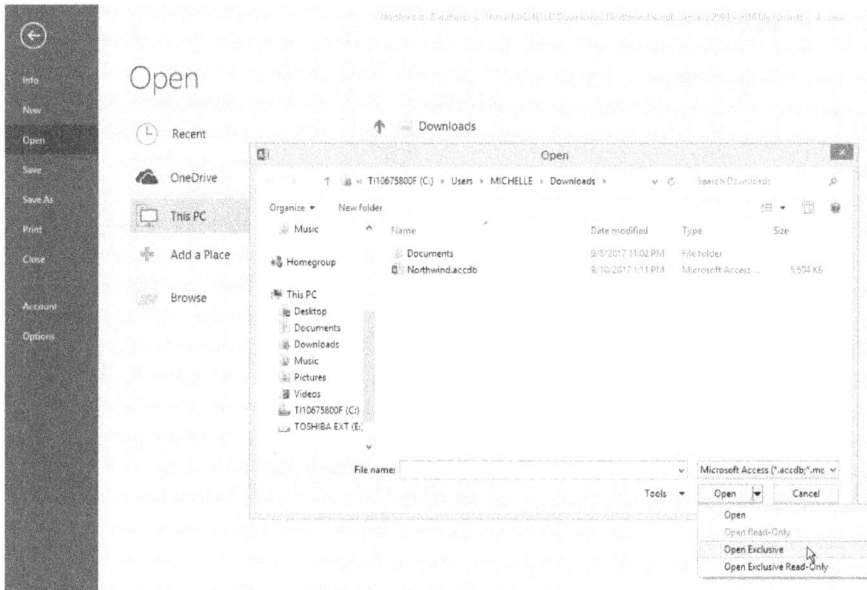

To encrypt a database with a password, use the following procedure.

Step 1: Select the File tab from the Ribbon to open the Backstage view.

Step 2: Select Encrypt with Password from the Info tab.

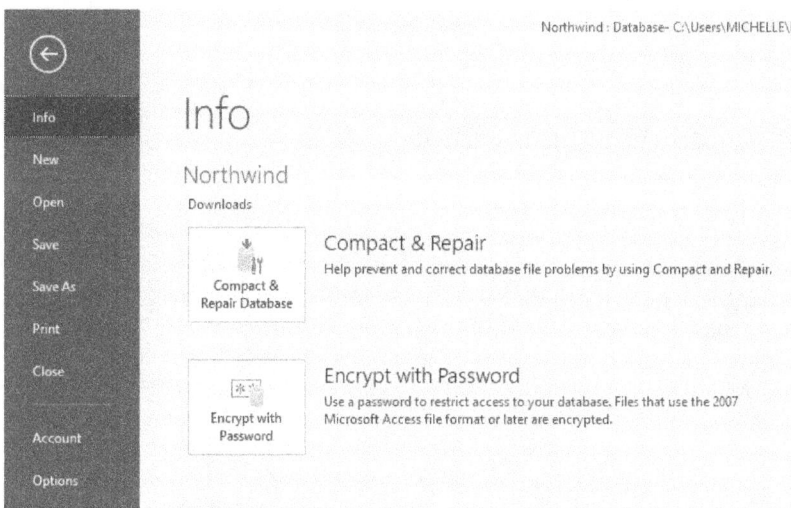

Step 3: In the Set Database Password dialog box, enter the Password you want to use. Remember to use a strong password.

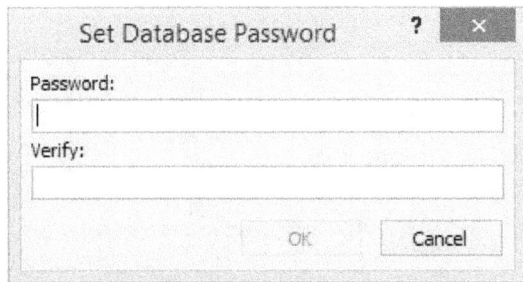

Step 4: Enter the Password again in the Verify field.

Step 5: Select OK.

To decrypt a database that has been password protected, use the following procedure.

Step 1: Select the File tab from the Ribbon to open the Backstage view.

Step 2: Select Decrypt Database from the Info tab.

Step 3: Enter the Password.

Using Save Object As

You can save a database object to use in another database.

To save the current object, use the following procedure.

Step 1: Open the table or object that you want to save.

Step 2: Select the File tab from the Ribbon to open the Backstage view.

Step 3: Select Save As.

Step 4: Select Save Object As.

Step 5: Select the Database Object As.

Step 6: Select Save As.

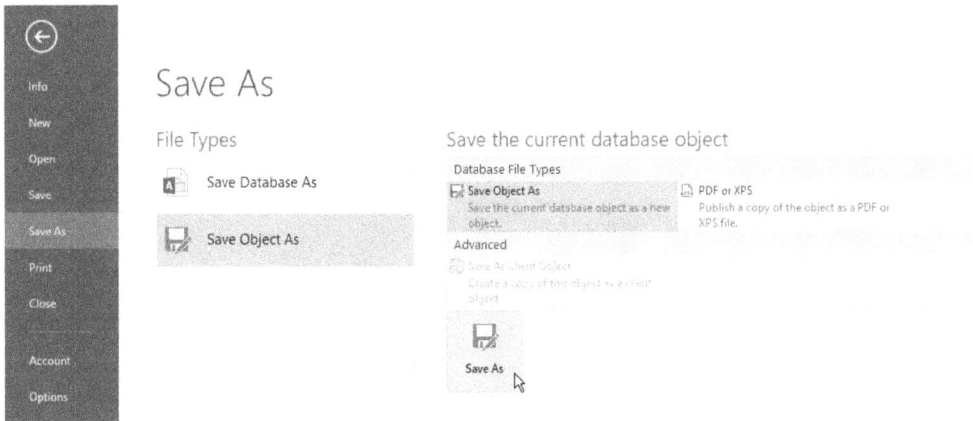

Step 7: In the Save As dialog box, enter a name for the table in the Save To field.

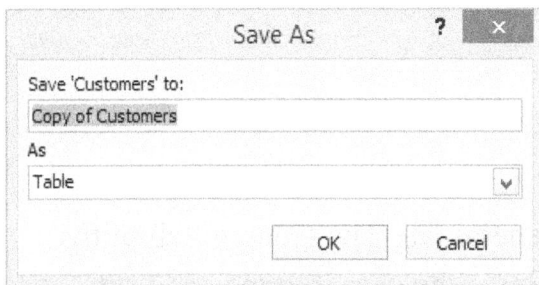

Step 8: Select the type of object from the As drop down list, if the appropriate object is not already selected.

Step 9: Select OK.

Setting Access Options

The Options dialog box allows you to customize the user interface.

To open the Options dialog box, use the following procedure.

Step 1: Select the File tab to open the Backstage View.

Step 2: Select Options.

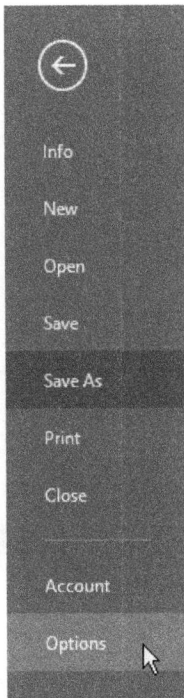

The Access Options dialog box opens to the General tab. The General tab of the Access Options dialog box allows you to customize the user interface and set options on creating new databases.

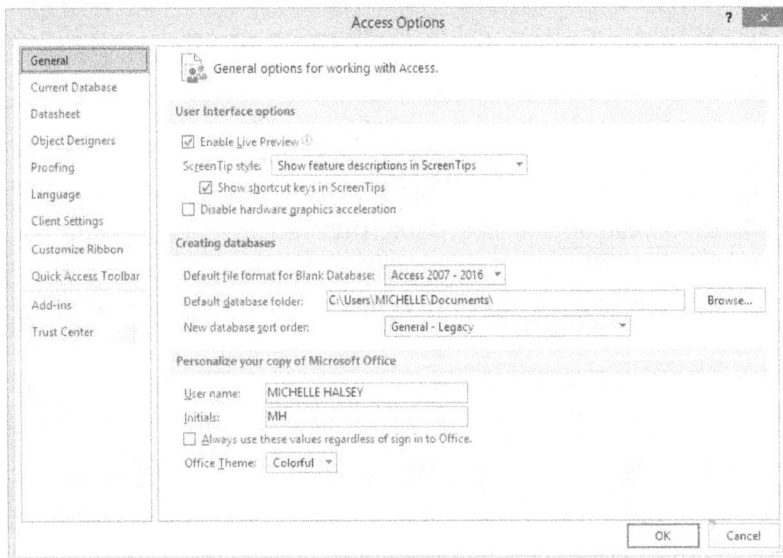

- Check the box to enable the selected features. Uncheck the boxes to disable those features.

- Under Creating databases, you can select the Default file format for a blank database. Choose either Access 2000, Access 2002-2003, or Access 2007-2016. You can also set a default location for creating new databases.
- You can also select a new database sort order.
- To personalize your copy of Access, enter your User Name and your Initials.

Select the Current Database tab to view those options.

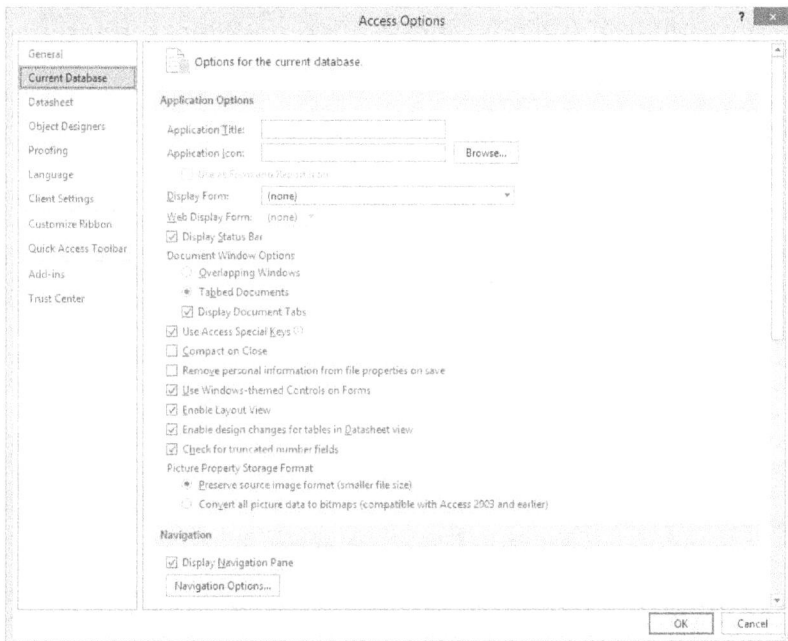

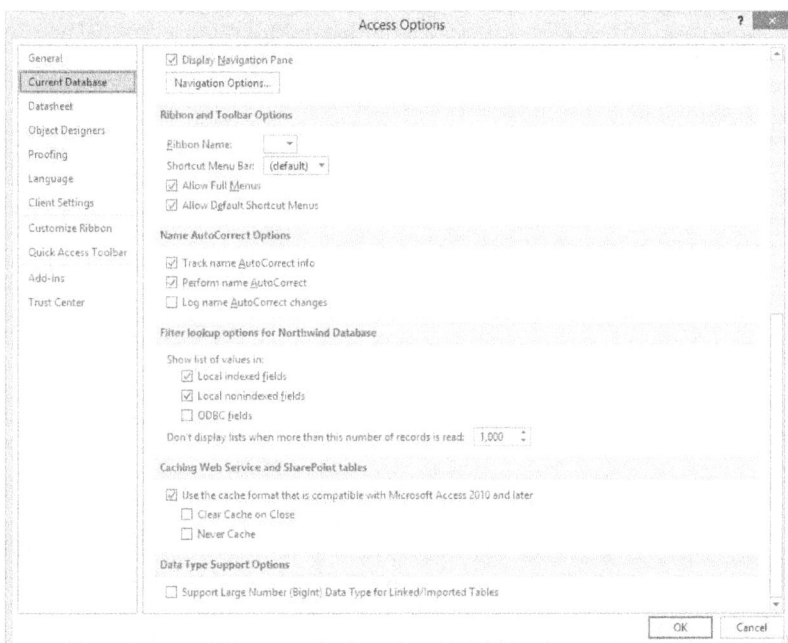

- You can customize the title and icon for the database. The title shows at the top of the database window.
- You can set a default Display Form or Web form by selecting an option from the drop-down list. If your database does not include forms, there will not be any options available.
- You can control whether the Status bar is displayed. You can also change the options for how the windows are displayed.
- Check the boxes for the additional options to enable them. Uncheck them to disable them.
- Choose a Picture Property Storage Format to control the compatibility of pictures used in the database.

Scroll down in the Current Database tab to see additional options. The Current Database tab includes the following other settings:

- Navigation
- Ribbon and Toolbar
- Name AutoCorrect
- Filter Lookup Options
- Caching Web Service and SharePoint Tables
- Data Type Support Options

These settings allow you to control additional options when you open this database.

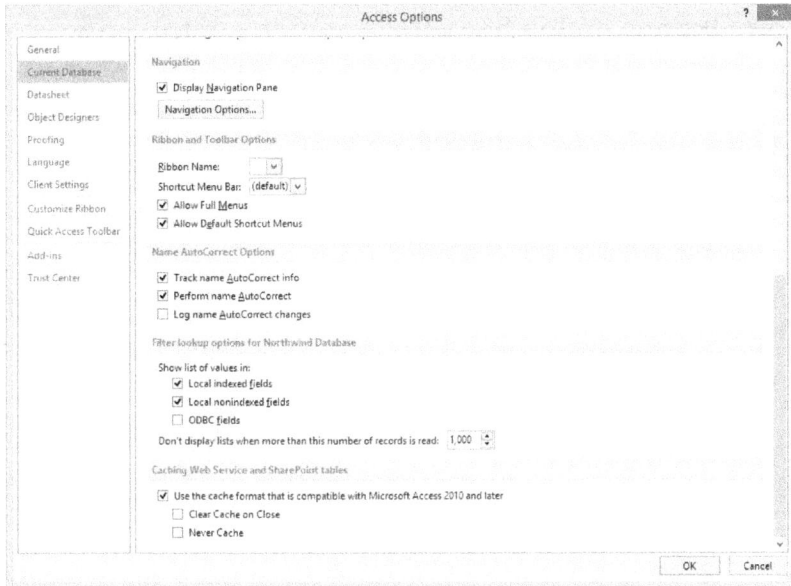

Check the box to indicate that you want the Navigation pane to show in this database. To customize the Navigation options, select Navigation options.

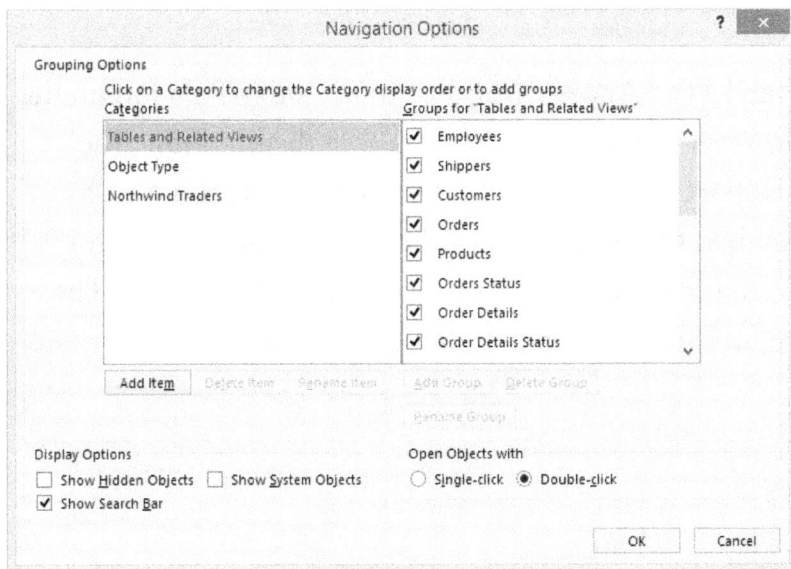

- You can select an item on the left list and Rename or Remove an Item. You can also Add Item.
- To organize the Navigation items into groups, use the right list. You can Add a Group, Delete a Group, or Rename a Group.
- Finally, you can enable additional display options and indicate how objects are opened.
- Select OK when you have finished setting Navigation options.

The Datasheet tab on the Access Options dialog box allows you to control how gridlines, cell effects, and fonts are shown in the database.

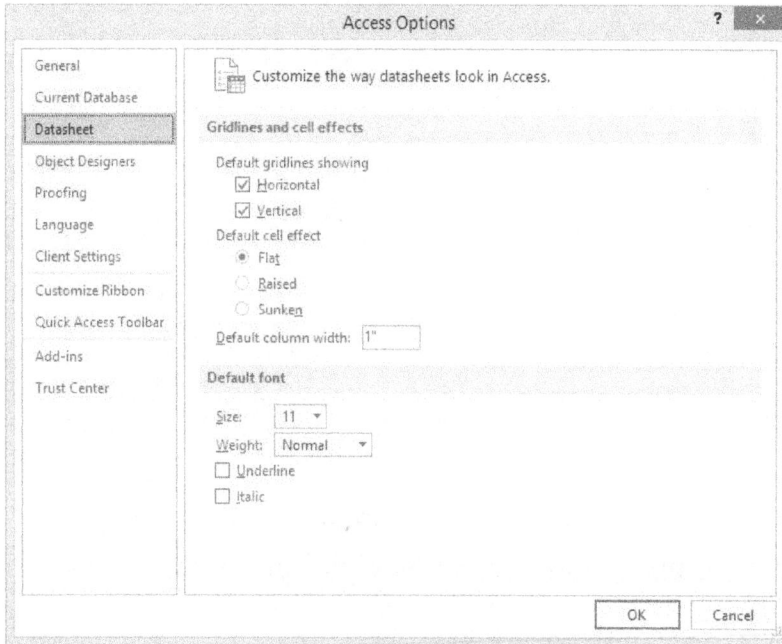

Review the Object Designers tab in the Access Options dialog box. This tab on the Access Options dialog box allows you to customize the Table design view, the Query design view, the Form/Report design view, and the Error Checking in form and Report design view. The Object Designers tab of the Access Options dialog box allows you to change the default settings for design of database objects. These settings do not apply if you are looking at objects in table datasheet and layout views.

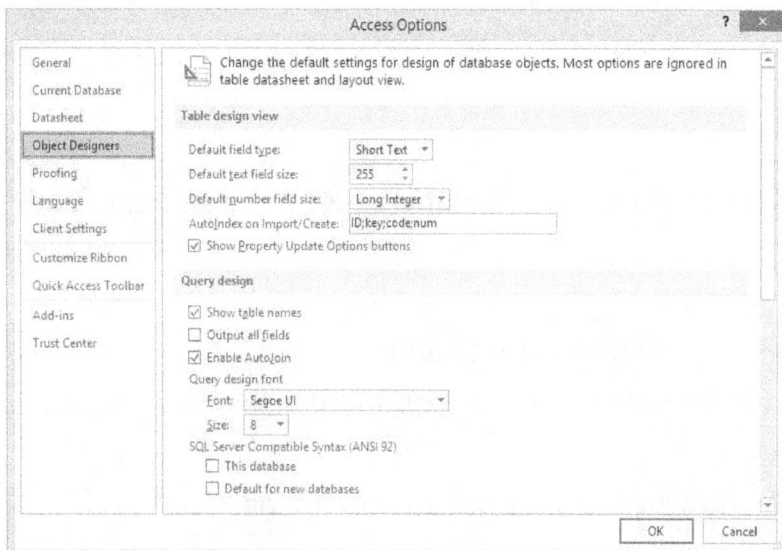

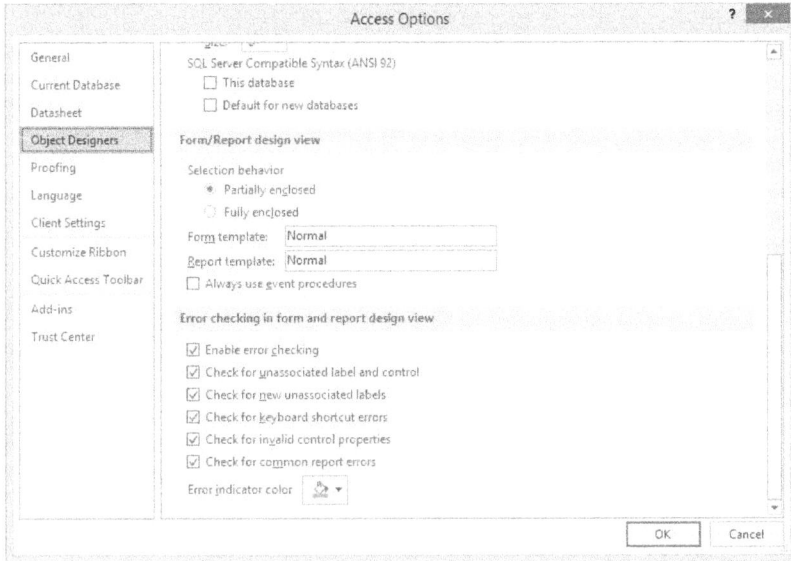

Review the Proofing tab in the Access Options dialog box. This tab on the Access Options dialog box allows you to control how Access corrects text as you type and how your spelling is corrected.

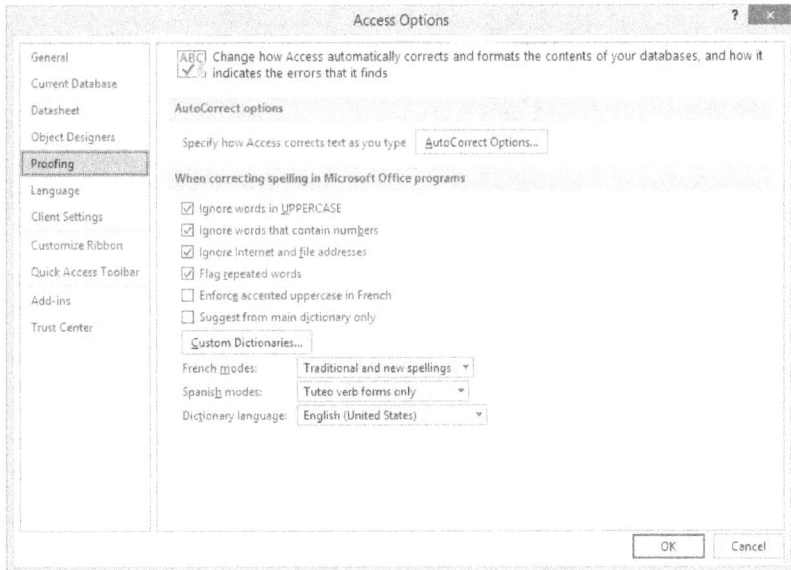

Select the AutoCorrect Options button to open the AutoCorrect dialog box.

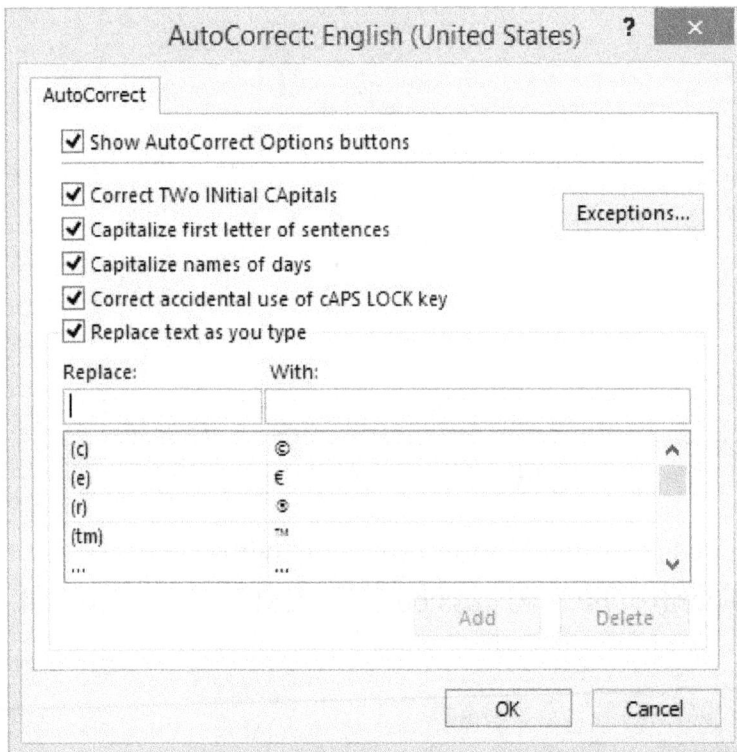

Select the Custom Dictionaries button to open the Custom Dictionaries dialog box.

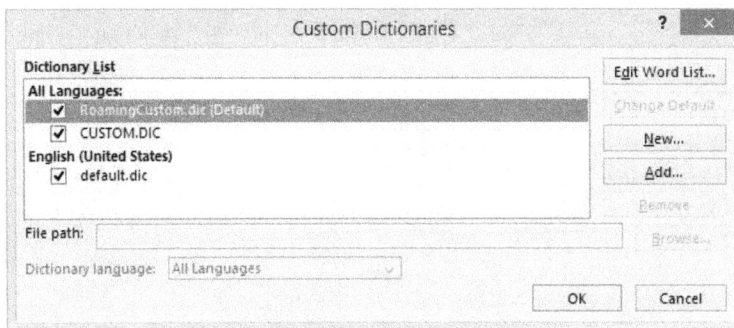

Review the Language tab in the Access Options dialog box. This tab on the Access Options dialog box allows you to control which language is used for proofing and for display and help.

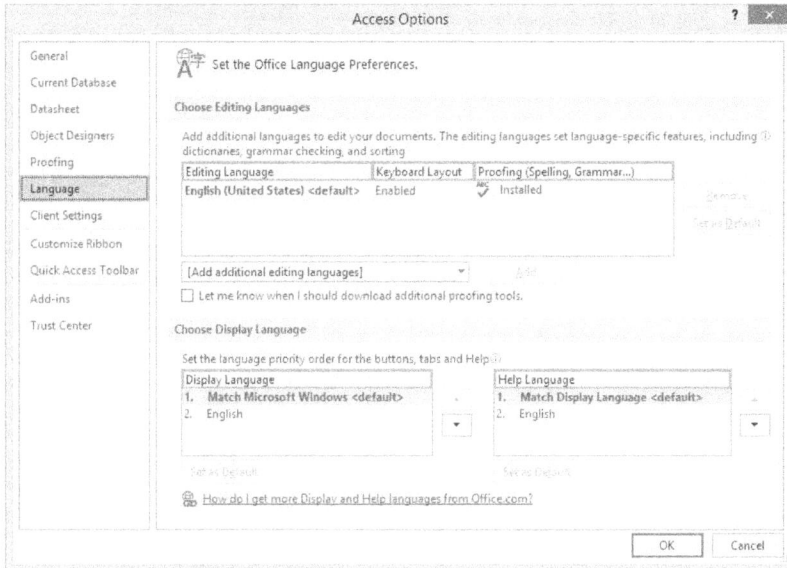

The Client Settings tab on the Access Options dialog box allows you to control how Access behaves when you are editing in a database. It also controls Display,

Printing, General and Advanced options related to your use of Access on the current computer. You can also change the default theme.

The Display options control how Access features are displayed on this computer. The Printing options control the margins when printing database items.

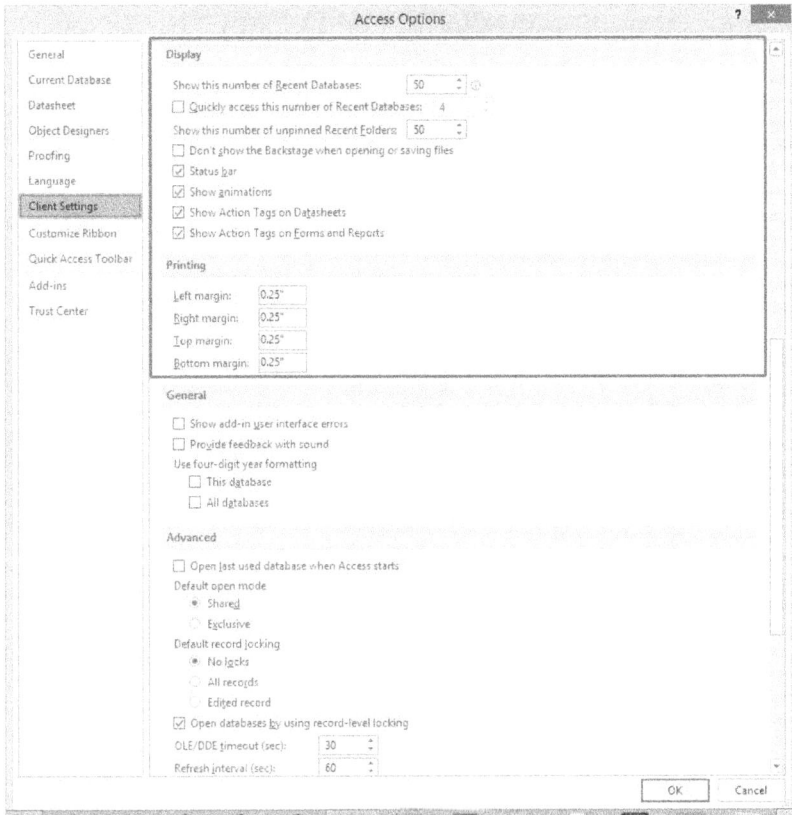

The General options allow you to enable or disable additional features. The Advanced options control features like the default open mode, default record locking, and different intervals and encryption methods. The Default Theme sets the theme used by Access on start up.

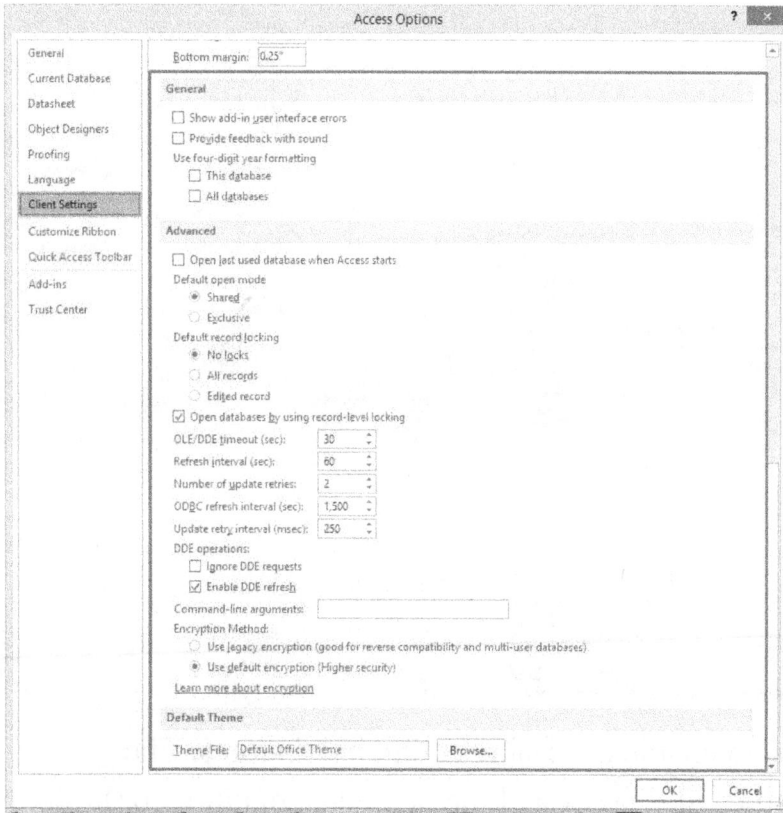

Chapter 23: Sharing Data Using Apps

Apps are a new type of database in Access 2016 that you can easily build and share with others in a web browser. This chapter will explain how to create a new app with a template. You will also use a table template to build your app. Then, you will learn how to launch your app and enter data. We will take a closer look at the app layout. Finally, you will learn how to upload your changes.

Creating a New App Using a Template

The quickest way to get a new app running is to use a template.

To create an app, use the following procedure.

Step 1: If you have another database open, select the File tab from the Ribbon. Select New. If you have just started Access, the templates are already shown on the startup page.

Step 2: Select an icon from the featured templates. The icons with a globe are for creating apps. In this chapter, we will use the Contacts example.

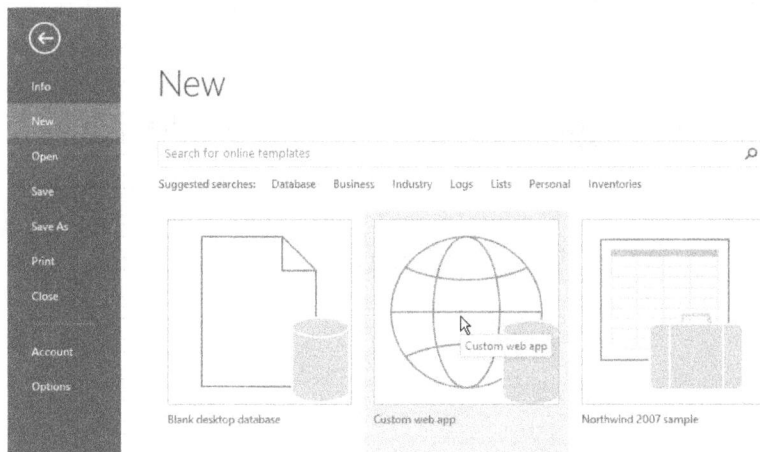

Note: If you do not see a globe, then search for "custom web app".

Step 3: Access displays information about the template. You can use the right or left arrows (in the gray shaded area) to scroll to descriptions of the other templates.

Step 4: Enter the App Name.

Step 5: Select the Web Location where you would like to store your app from the Available Locations list. Or you can enter a new Web Location.

Step 6: Select Create.

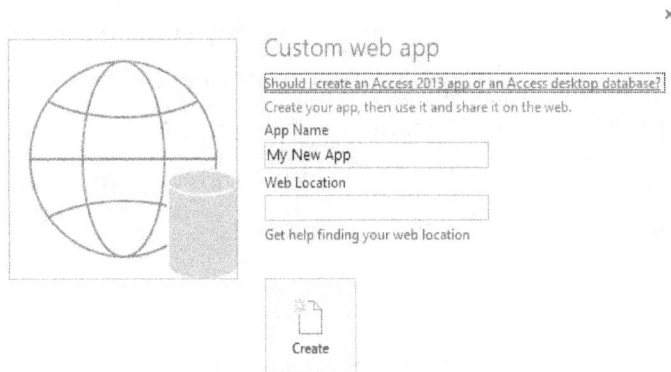

Your app has been created.

Selecting a Table Template

After you create your app, you still need to make sure you have the right tables to track your data. Use one of the Access templates to make the job easy.

To add a table, use the following procedure.

Step 1: If the Add Tables page is not already showing, select Add Tables from the bottom of the App tab on the left.

Step 2: On the Add Tables page, select one of the links for a Suggested Search or enter a search term to help you find a template for the type of table you want to include.

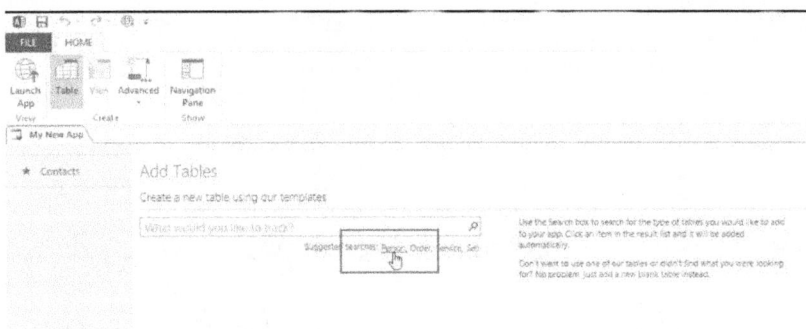

Step 3: From the list that appears under the search box, select the table you want to add. Access adds it immediately.

In this example, we have added both a Customers table and an Employees table. You can see the tables in the My New App Navigation area.

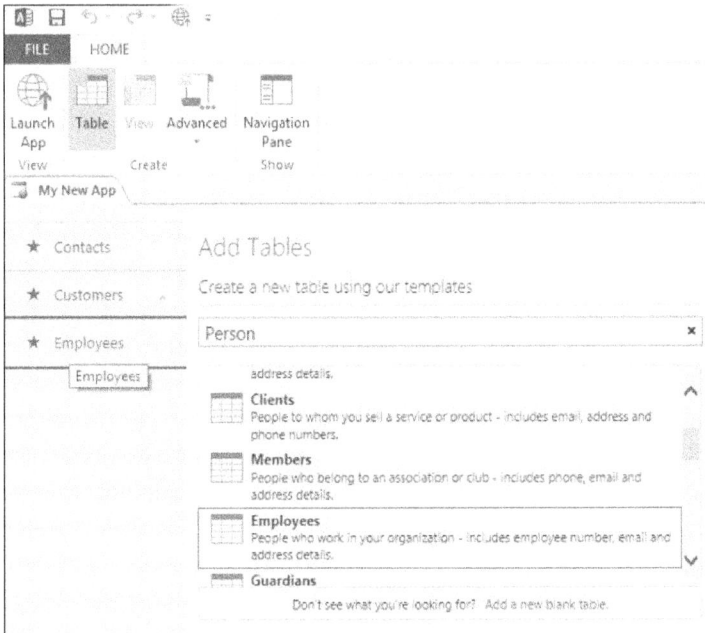

Creating a Table By Importing Data

Instead of, or in addition to creating new tables, you can also create a table by importing existing data.

To create a table by importing data, use the following procedure.

Step 1: If the Add Tables page is not already showing, select Add Tables from the bottom of the App tab on the left.

Step 2: On the Add Tables page, select the data source where your data is located. In this example, we will select Excel.

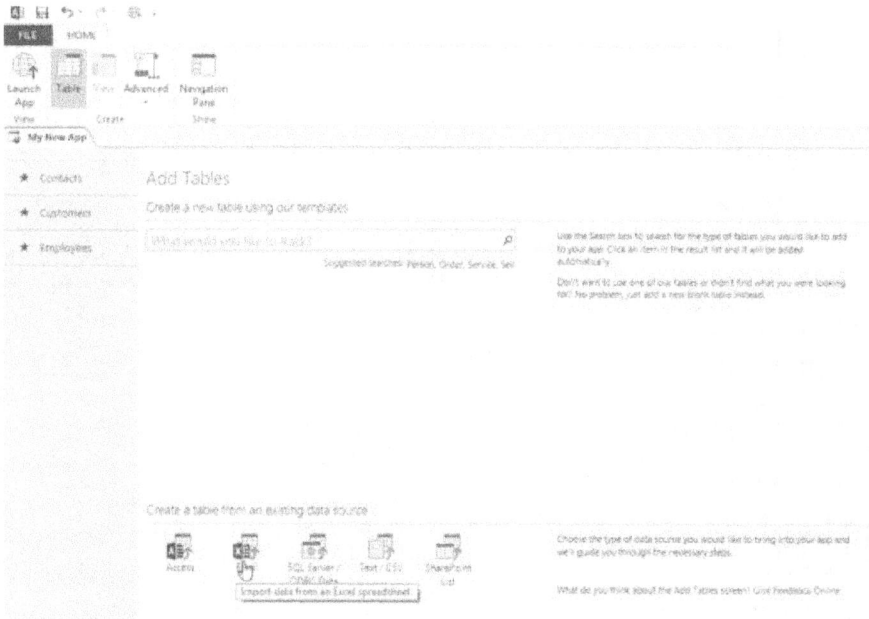

Step 3: In the Get External Data dialog box, select Browse to locate the file.

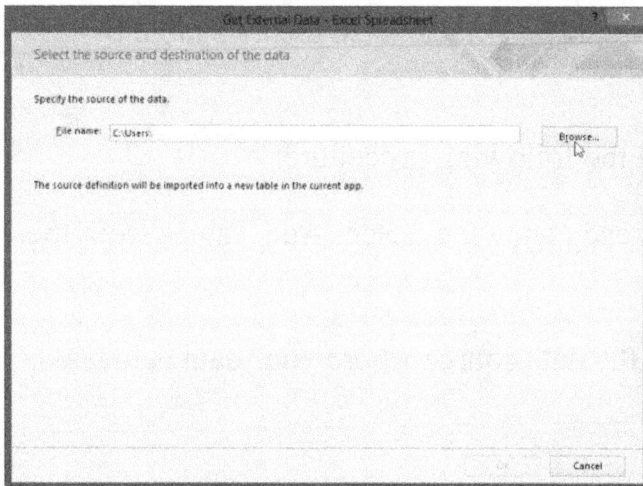

Step 4: In the File Open dialog box, navigate to the location of the file that contains your data. Select Open.

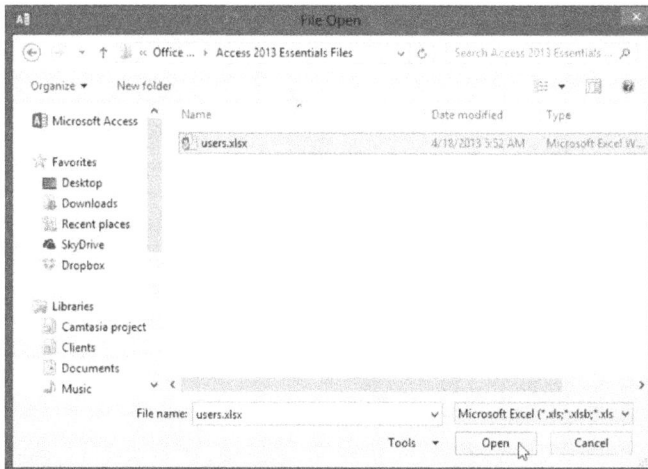

Step 5: In the Get External Data dialog box, select OK.

Step 6: Access displays the Import Spreadsheet Wizard dialog box to help you review your data. Check the First Row Contains Column Headings box if applicable. Select Next.

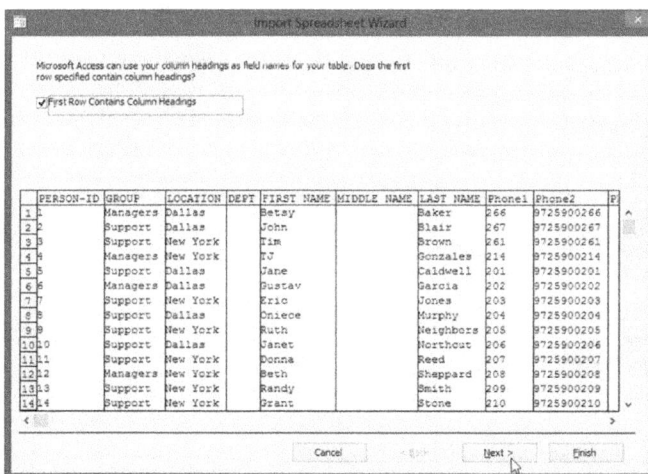

Step 7: The next screen allows you to review each column heading. You can change the Field Name by entering new text. You can change the Data Type by selecting a new option from the drop-down list. You can also check the Do Not Import Field box if you want to skip this field. Repeat for each column heading by clicking anywhere on each column. Select Next.

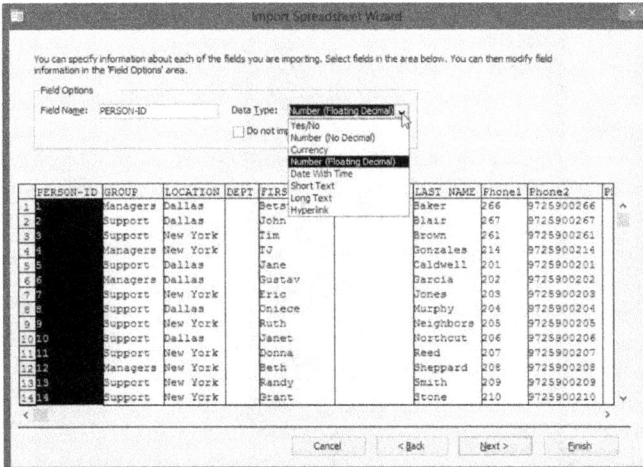

Step 8: The final screen of the Import Spreadsheet Wizard allows you to review the name of the table. Enter new text in the Import to Table field to change the name.

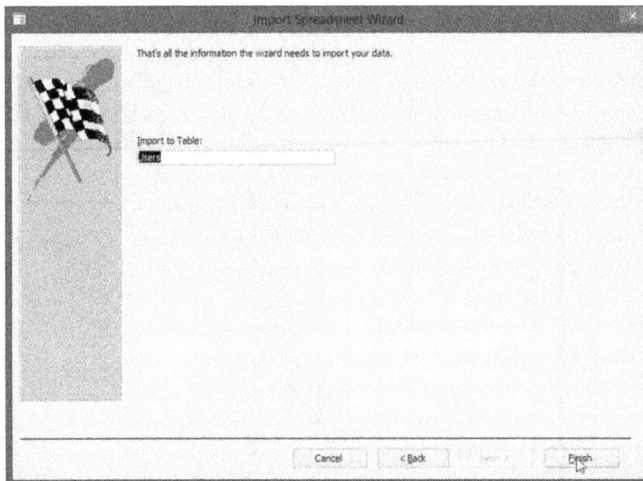

Step 9: Select Finish.

Step 10: Select Close on the Import Operation Finished screen.

Launching the App

Your app is ready to use once you launch it. In the remainder of the course, you will learn skills that you can use if you choose to customize your app.

To launch the app, use the following procedure.

Step 1; On the Home tab of the Ribbon, select Launch App.

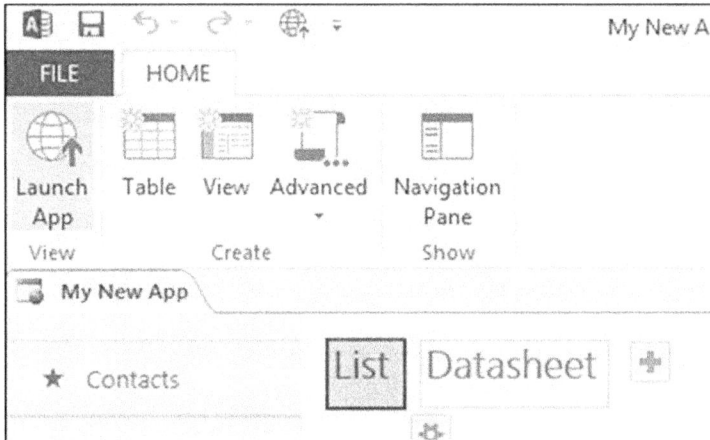

The app launches in your default web browser.

You can start entering data on the tables or view data for any tables that were created by importing data.

Chapter 24: Working in Your App

Now that you have launched your app, let's take a closer look at it. We will look at how to use the Search Box and the Action Bar. You will also learn about automatically generated controls. Next, we will look at viewing related items and using AutoComplete to look up a related item. Finally, we will look at grouping and summarizing data.

Using the Search Box

The Search box available in the List view allows you to find records that match criteria you enter.

To search for an item in the app, use the following procedure.

Step 1: Make sure that you have selected the correct table (on the left side of the screen). In this example, we will use the Users table that we imported in the previous chapter.

Step 2: Make sure that you have selected the List view from the link at the top.

Step 3: Enter an item in the search box to filter the list. You can enter text from any field. In this example, we will search for all users in the Dallas location. Press the Enter key or select the magnifying glass to perform the search. The list now only shows records that match the search criteria.

Step 4: Click the X in the search box to cancel the filter.

Using the Action Bar for Predefined Actions

The icons to the right of the search box allow you to:

⊕	Add a new record
🗑	Delete the current record
✏	Edit the current record
💾	Save a record you just entered
✗	Cancel a record you just entered

Let's practice the actions in the List view.

To add a record, use the following procedure.

Step 1: Select the Add icon. Notice that the list now includes an entry labeled (New).

Step 2: Enter the data. Press Tab to move to the next field.

Step 3: Select the Save icon to save your changes. If you do not want to keep the record, select the Cancel icon.

To edit a record, use the following procedure.

Step 1: Make sure that the record you want to edit is showing.

Step 2: Select the Edit icon.

Step 3: Correct the fields you want to change by entering the new information. Press Tab to move to the next field. The information is highlighted to make it easy to replace.

Step 4: Select the Save icon to save your changes. If you do not want to keep the record, select the Cancel icon.

To delete a record, use the following procedure.

Step 1: Make sure that the record you want to delete is showing.

Step 2: Select the Delete icon.

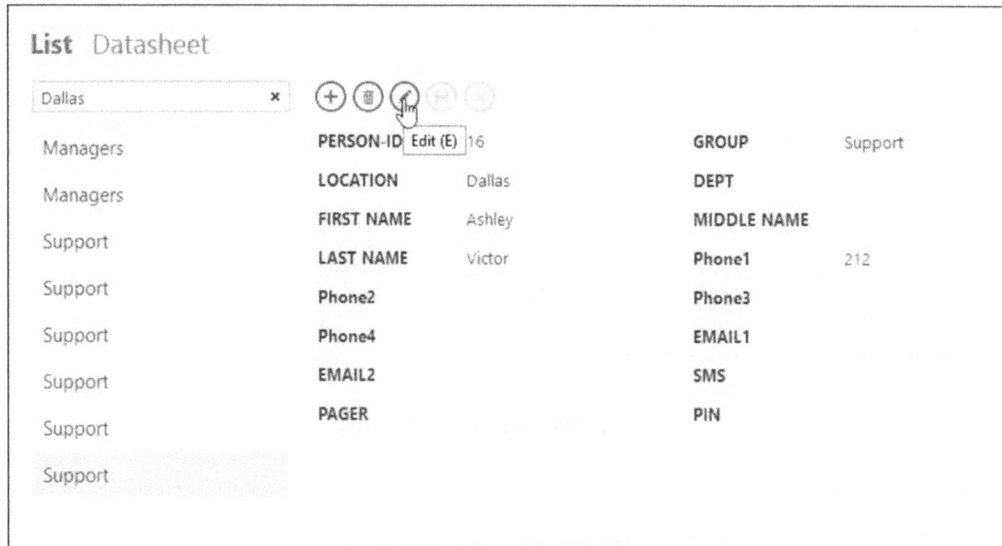

Step 3: In the Delete Confirmation window, select Yes.

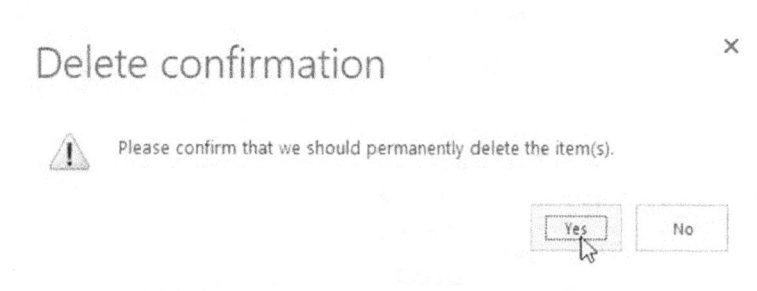

In the Datasheet view, the only icons are Add and Delete. You can edit a record simply by clicking in the field you want to change and entering the new information. Press Enter to save the changes.

Working with Related Items and AutoComplete

Access 2016 makes working with related data easy used related item controls and autocomplete controls.

Set up the app for this activity, use the following procedure.

Step 1: Create and launch a Task Management app.

Step 2: Enter and save one or more records in the Employees table. The First and/or Last Name is the most important information for this activity – not all the fields need to contain information.

To use the Autocomplete control, use the following procedure.

Step 1: Switch to the Tasks table.

Step 2: Enter a new task record. Complete at least the Task Title. The other fields can remain blank.

Step 3: Press TAB (or use the mouse to click on) the Assigned To field.

Step 4: Begin typing one or the first or last names you entered in the Employee table.

Notice that the app displays matches from the Employees table in a drop-down list as soon as you start typing. Click on the item you want to enter.

Step 5: Save the task.

Now let's look at the related items control, use the following procedure.

Step 1: Switch back to the Employees table.

Step 2: Scroll down to the bottom. Notice that the Tasks section that lists any tasks assigned to the selected employee.

List Datasheet By Group

Filter the list...

State/Province   ZIP/Postal Code
Country/Region   Office Location
Department   Date of Hire
Date of Birth   Emergency Con...
Emergency Con...   Emergency Con...
Web Page   Notes
Group   Contract

Tasks

Task Title	Status	Due Date	Description
Finish Module 5	In Progress		

Add Tasks

Step 3: Select the task to open a popup window of that task record.

Tasks   ✕

Task Title   Finish Module 5   **Priority**   2 - Medium
Description   **Start Date**
                                     Due Date
                                     Percent Compl... 0.00%
Hours   **Status**   In Progress
Active   ☑   **Assigned To**   Michelle Halsey
          Hours

Step 4: Select the Add Tasks link at the bottom to add tasks for the selected employee.

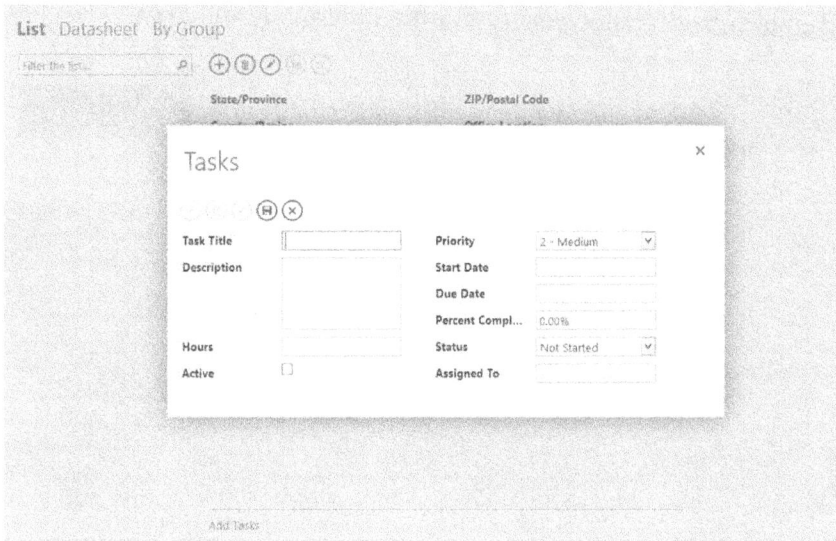

Grouping Data

Depending on what kind of tables your app includes, Access may have automatically added a Summary view. The Summary view allows you to view your data by selected groups.

To view information by groups, use the following procedure.

Step 1: Select the Summary view from the list of views at the top. Depending on your table, it may be labeled By Group or By Status.

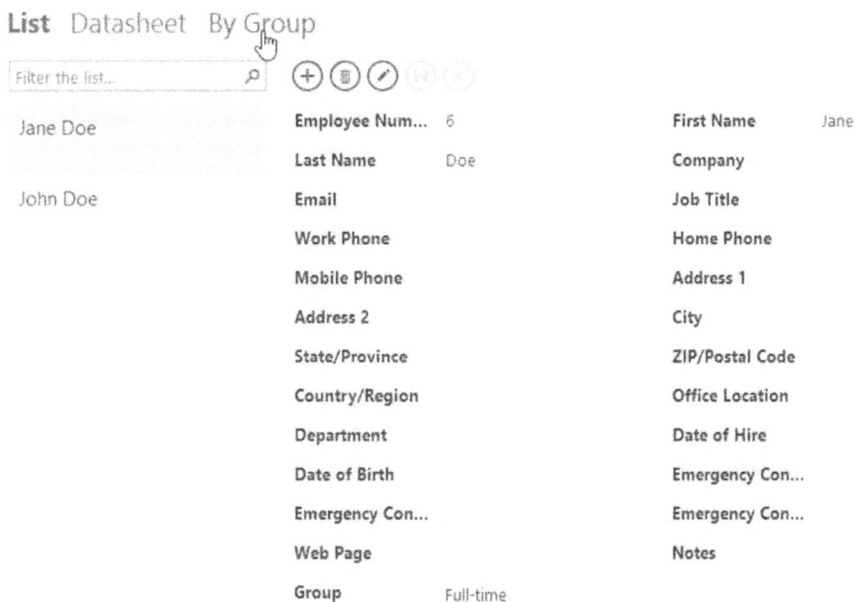

Step 2: All the groups for the selected table are shown in the list on the left.

Step 3: The right side shows the records for the selected group. Select an item in the list to open that record in a popup window.

List Datasheet **By Group**				
Filter the list... 🔎	Display Name	Company	Job Title	Work Phone
Contract (1)			Trainer	
Full-time (1)				
Part-time (1)				

Chapter 25: Creating a Custom App

Access 2016 has many ready-to-use apps. You may have even experimented with modifying an app so that it can better fit your needs. But sometimes, it is better to just start from scratch to get the app that you need. In this chapter, you will go through the process of creating your own app using the Custom web app starter, a combination of template tables and blank tables, and importing tables from Access 2010.

Creating a Custom Web App

To start your custom web app, you will need to use the Custom Web App starter in Access.

To start a custom web app, use the following procedure.

Step 1: Select the File tab from the Ribbon.

Step 2: Select New.

Step 3: Select Custom Web App.

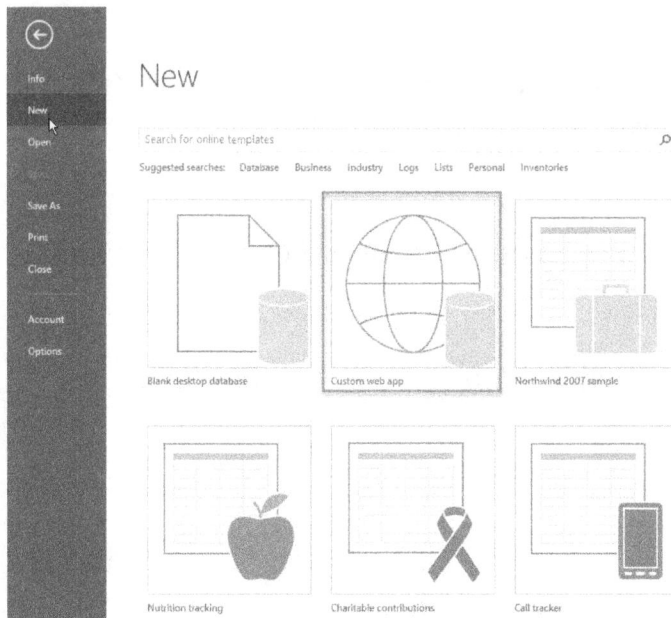

Step 4: Access displays information about the template. You can use the right or left arrows (in the gray shaded area) to scroll to descriptions of the other templates.

Step 5: Enter the App Name.

Step 6: Select the Web Location where you would like to store your app from the Available Locations list. Or you can enter a new Web Location.

Step 7: Select Create.

Access "talks" with the server for a few moments to set up the app.

Adding a Template Table

Your custom app has does not have any tables yet. Use one of the Access template tables to start building your database.

To add a table, use the following procedure.

Step 1: On the Add Tables page, select one of the links for a Suggested Search or enter a search term to help you find a template for the type of table you want to include.

Step 2: From the list that appears under the search box, select the table you want to add. Access adds it immediately.

In this example, we have added a Clients table. You can see the tables in the App Navigation area.

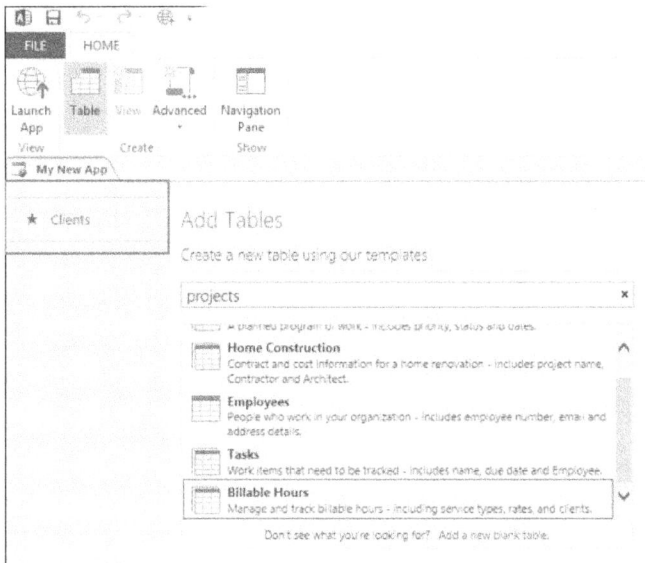

We have also used the Billable Hours template, which added hours, rates, and employees tables.

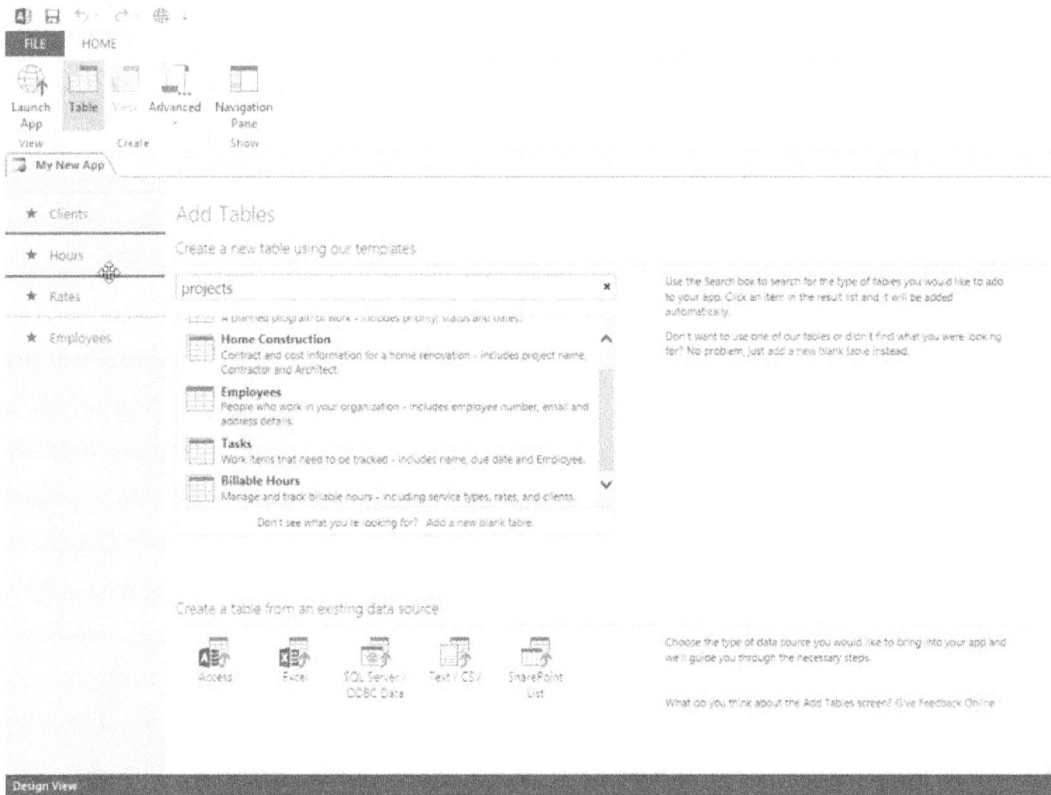

Adding a Blank Table

If you do not find a template for the type of table you want to add, you can use a blank table and create the fields and properties you need.

To add a blank table, use the following procedure.

Step 1: Select Add Blank Table from the Add Tables page.

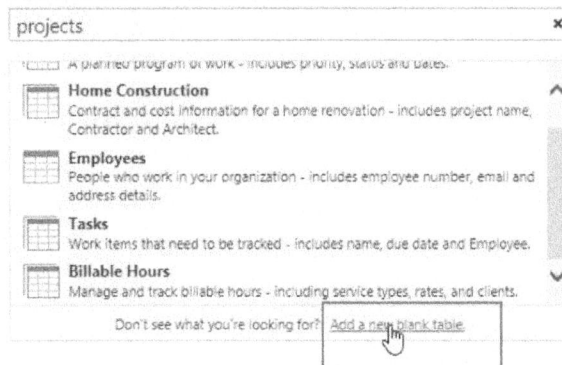

Step 2: Access opens the table in Design view so that you can develop your table quickly using the table design tools.

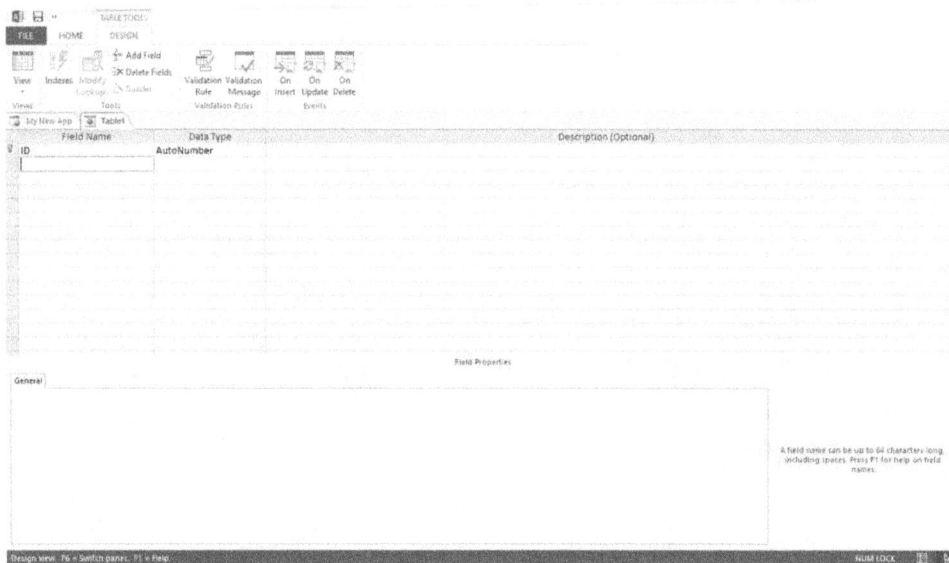

Step 3: For each field, you want to include in the table, enter the following information:

- Field Name
- Data Type
- Description (optional)

Step 4: If you need to change additional field properties, select the field you want to change and adjust the Field Properties in the section at the bottom of the screen.

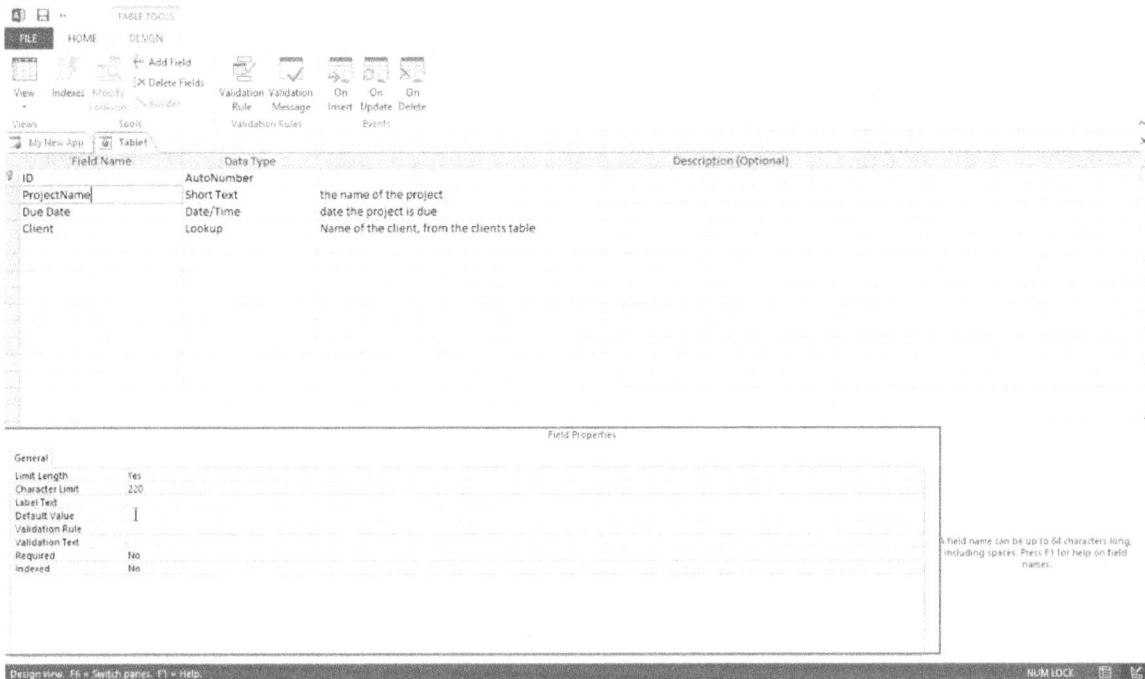

Step 5: When you have finished designing your table, make sure you save it. You can also save changes to the list view and the datasheet view as needed.

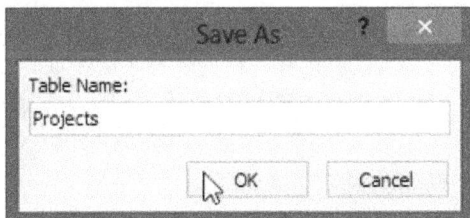

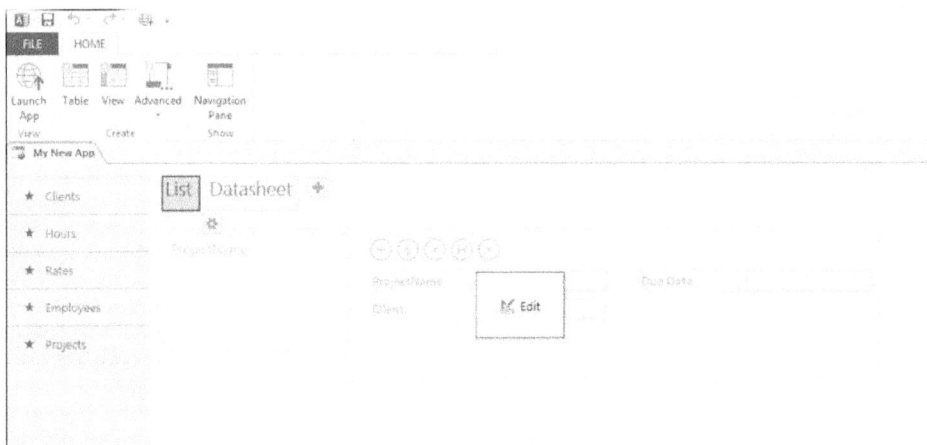

Importing Access 2010 Tables

Access 2010 databases cannot be updated to web apps. But you can import them to use as an app.

To import an Access 2010 table, use the following procedure.

Step 1: Select Add Tables at the bottom of the list of tables to open the Add Tables page.

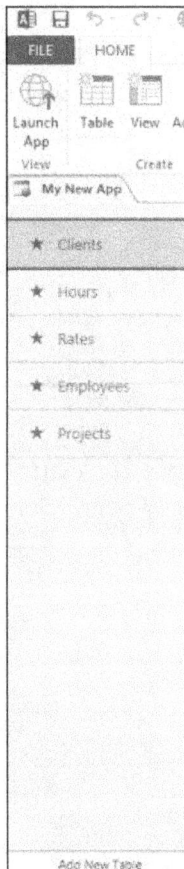

Step 2: From under the heading "Create a table from an existing data source," select Access.

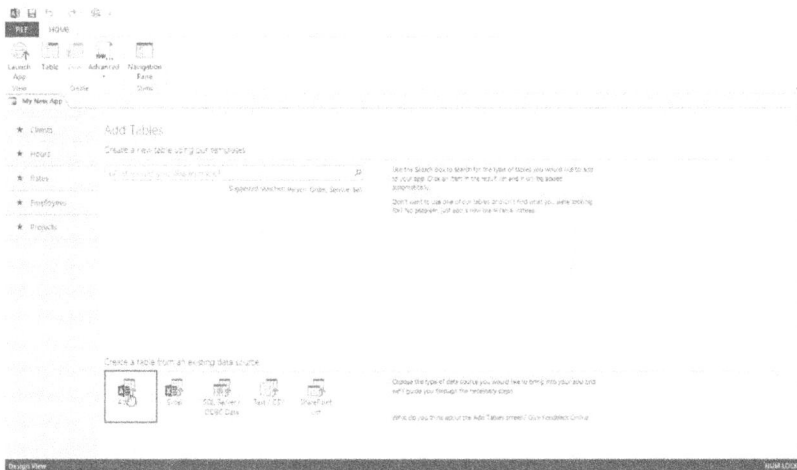

Step 3: In the Get External Data dialog box, select Browse to open the File Open dialog box.

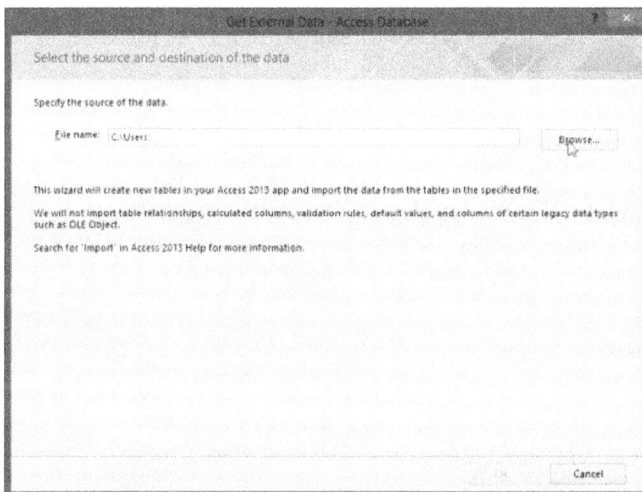

Step 4: From the File Open dialog box, navigate to the 2010 Access Database or table that you want to import. Select the file and select Open.

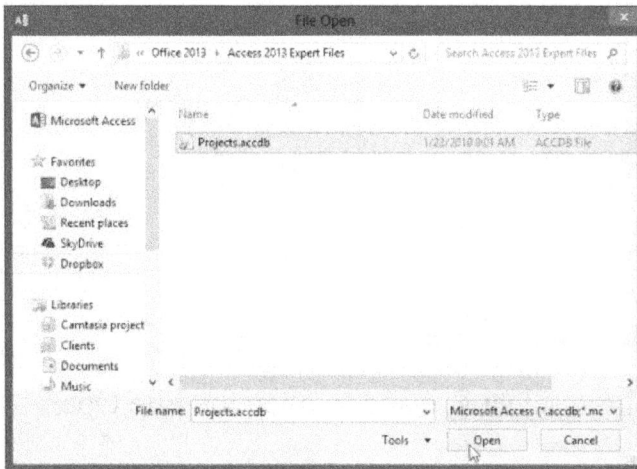

Step 5: In the Get External Data dialog box, select OK to have Access create new tables based on your selection.

Step 6: Access opens the Import Objects dialog box. Select the tables you want to import and select OK. You can also select Options to see additional import information.

Chapter 26: Customizing App Actions

When you are designing your app, you can add new actions to the Action bar. These actions can be customized using macros. Macros are like a simplified programming language that you write by building a list of actions you want Access to perform.

This chapter will help you create, edit, or delete custom actions.

Adding a Custom Action

Custom Actions are a way to customize your app even farther. You can add custom actions to any app, including apps you created from a template and apps you created from scratch.

To create a custom action, use the following procedure.

Step 1: Highlight the table in the list of tables on the left side for your app.

Step 2: Highlight the view that you want to modify.

Step 3: Select the settings icon to the right and just below the view name.

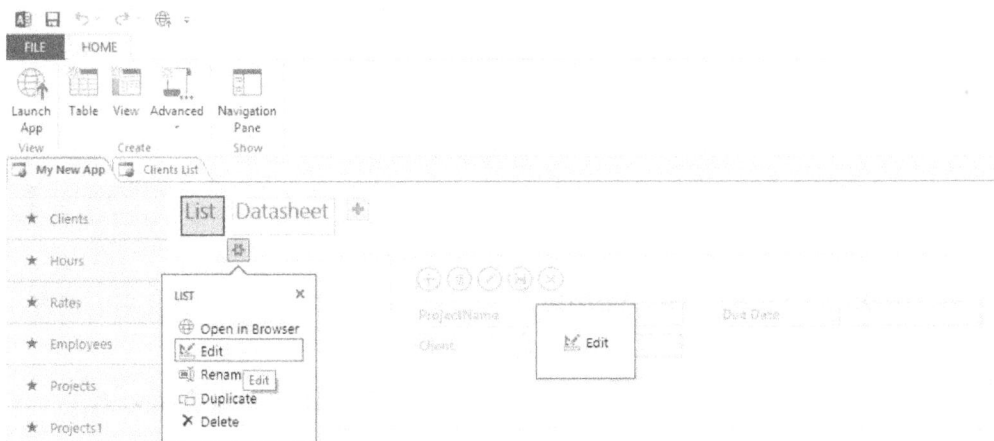

Step 4: Select Edit.

Step 5: Select the Add Custom Action icon.

Step 6: A new icon is immediately added. Select the Data icon to modify the action.

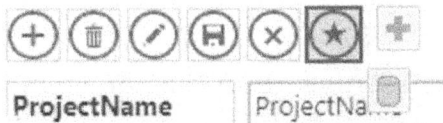

Step 7: In the Data dialog box, you will set the following properties:

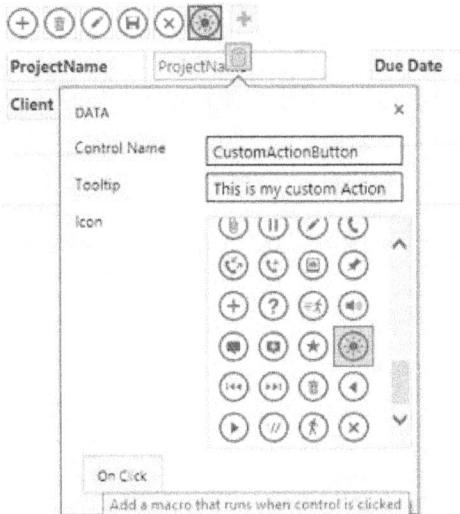

- Enter a Control Name for use within Access. This name will not show to users.
- Enter a Tooltip to explain your action item. Users will see this tooltip when they hover the mouse over the action item.
- Select an Icon from the gallery. Make sure to select one that is not currently being used in the same view, or you will have two different actions with the same icon.
- Select On Click to open a blank macro in Macro Design View.

Step 8: In the Macro Design view, select the first action you want the action button to perform from the drop-down list. You can also add macro actions from the

236

Action Catalog on the right. Just double-click on the action you want to include. Notice that if you click once on an action in the Action Catalogue, Access displays a definition at the bottom of the Action catalog pane.

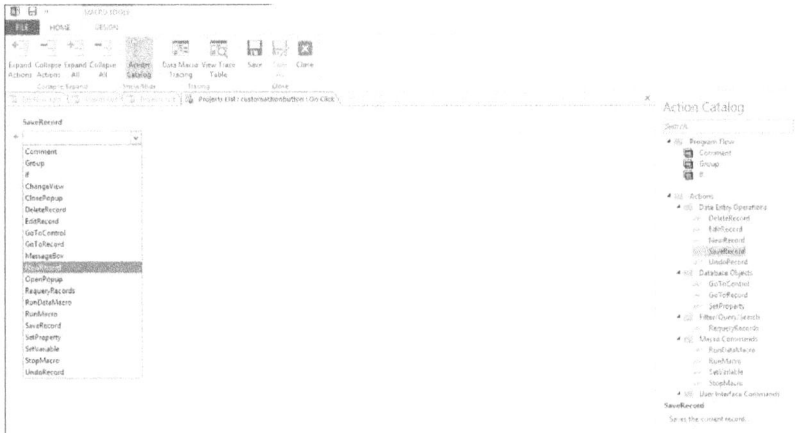

Step 9: If you select actions that require the arguments, fill in the argument information. We will discuss more about user interface macros in the next chapter.

Step 10: If you add multiple actions, you can use the arrow keys on the right to rearrange them. You can also click the X to delete an action you no longer need.

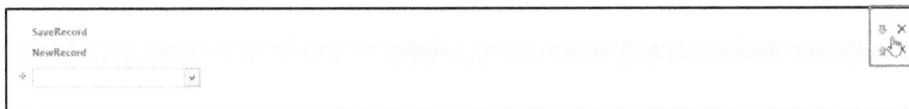

Step 11: Save your changes and close the macro design view.

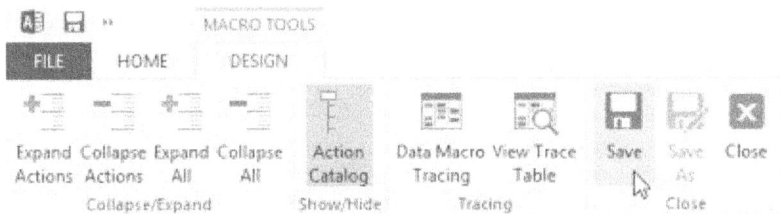

Step 12: In the Data dialog box, the On Click button is now green to indicate that you have added actions.

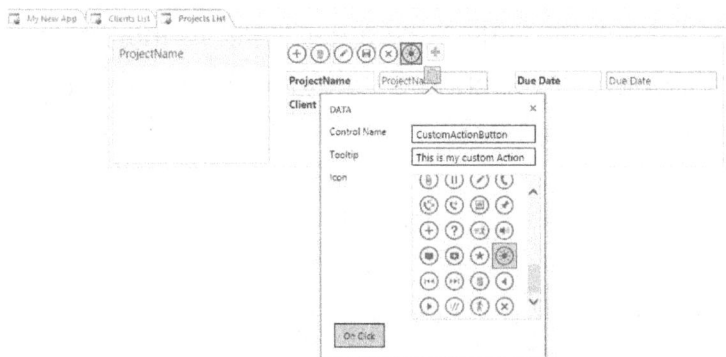

Step 13: Also make sure to save your database and launch the app to upload your changes to the browser.

Editing an Action

You can make changes to custom actions you have added to the action bar in the current view of your app.

To edit a custom action, use the following procedure.

Step 1: Highlight the table in the list of tables on the left side for your app.

Step 2: Highlight the view that you want to modify.

Step 3: Select the settings icon to the right and just below the view name.

Step 4: Select Edit.

Step 5: Select the icon for the action you want to modify.

Step 6: Select the Data icon to modify the action.

Step 7: In the Data dialog box, you can change the following properties:

- Enter a Control Name for use within Access. This name will not show to users.
- Enter a Tooltip to explain your action item. Users will see this tooltip when they hover the mouse over the action item.

- Select an Icon from the gallery. Make sure to select one that is not currently being used in the same view, or you will have two different actions with the same icon.
- Select On Click to open Macro Design View for your macro.

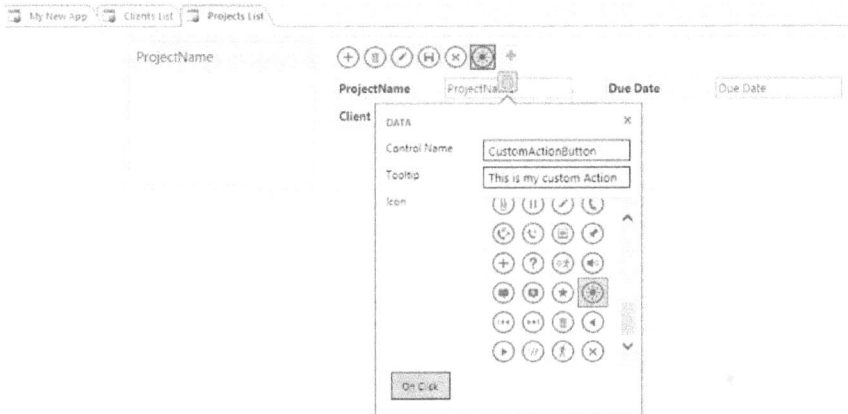

Step 8: In the Macro Design view, you can add actions either from the drop-down list or the Action Catalogue, you can edit arguments as needed, you can rearrange actions, or you can delete actions.

Step 9: Save your changes and close the macro design view.

Step 10: Also make sure to save your database and launch the app to upload your changes to the browser.

Deleting a Custom Action

If you no longer need a custom action you can delete it just as you can "delete" predefined actions (which means they are no longer shown on that view).

Step 1: Highlight the custom action icon you want to remove.

Step 2: Press the Delete key or the Backspace key.

The action is immediately removed. There is no confirmation, but Undo is available if you delete an action by accident.

Step 1: Also make sure to save your database and launch the app to upload your changes to the browser.

Chapter 27: Using App Views

Views are designed for Access 2016 apps to help you find your data fast. In this chapter, you will learn how to add your own view. You will also learn how to rename the view and duplicate an existing view. This chapter explains how to add a popup view so that you can access that view from a field with some types of controls. Finally, we will look at how to delete a view you no longer need.

Adding a New View

Access creates a List Details view and a Datasheet view for each table in your app. The List Details view includes a search box by default, to help you filter your data. The filtering in Datasheet view works more like it would in a spreadsheet program (just click on the column heading to see the options).

Some tables in your apps include Summary views. Summary views group items by a value. The Search box appears here as well, but it is limited to searching to just the list below it. The search capabilities do not extend to the list on the right.

You can create additional views that be customized in different ways.

To add a new view, use the following procedure.

Step 1: Open the app in Access.

Step 2: Select the table for which you want to create the view.

Step 3: Select Add New View (the plus sign).

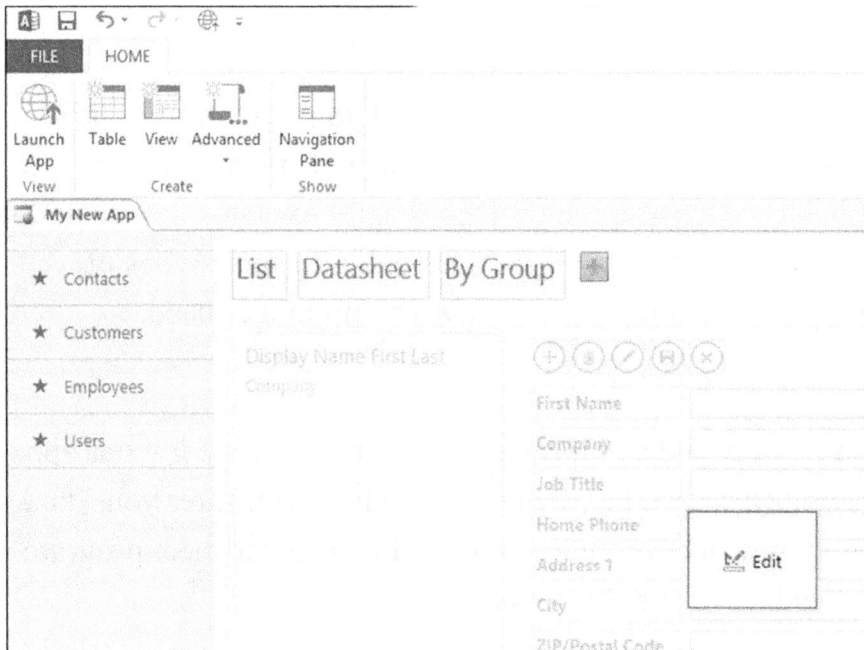

Step 4: In the Add New View dialog box, enter a View Name.

Step 5: Select the View Type from the drop-down list to indicate whether you want a list view, a datasheet view, a summary view or a blank view to customize from scratch.

Step 6: Select Add New View.

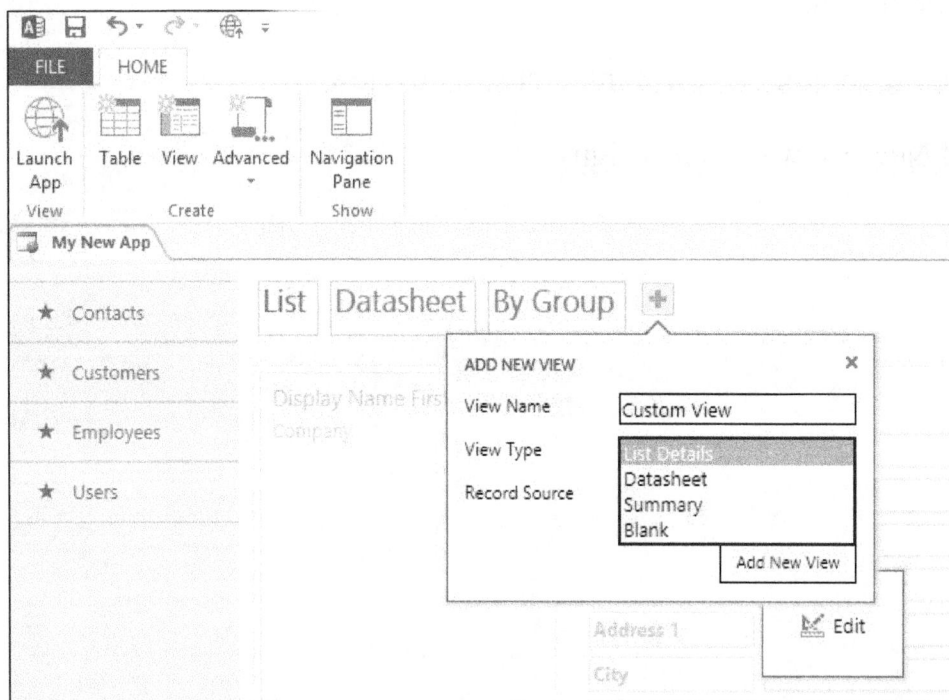

Step 7: Select Launch App to send the changes to the app in the browser.

Editing a View

You can edit your views to customize them. In this lesson, we will look at renaming and duplicating your view.

To edit a view, use the following procedure.

Step 1: Open the app in Access.

Step 2: Select the table and view that you want to edit.

Step 3: Select the Settings icon just to the left and below the View name.

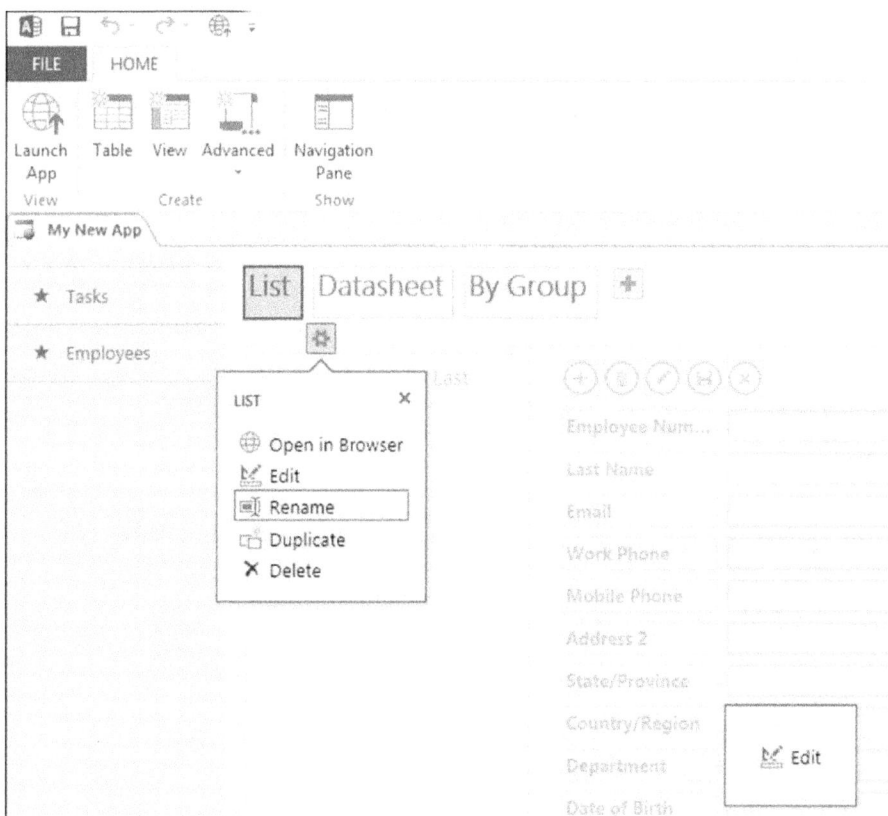

Step 4: Select Rename to rename the view. Enter the new Name and press Enter.

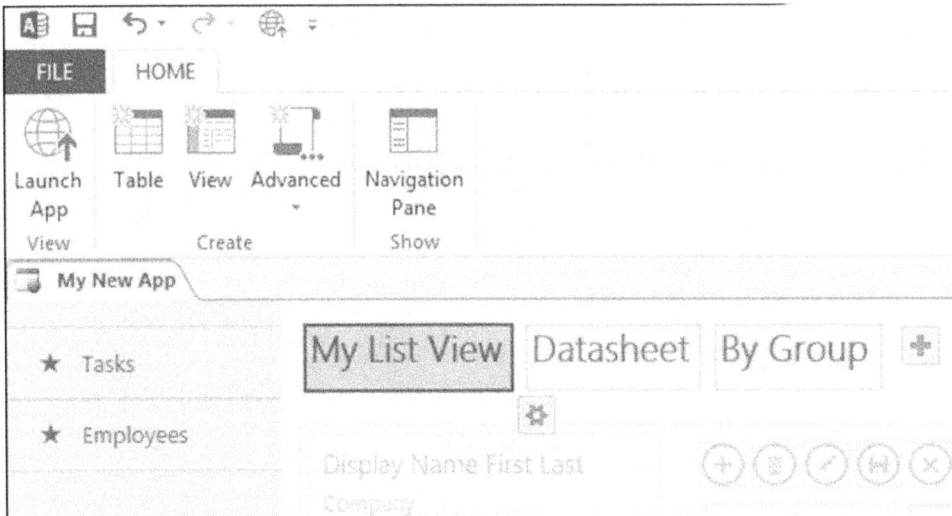

Step 5: Select Duplicate to create a new view based on the selected view. In the Duplicate View dialog box, enter the Name for the new view. Select the table where the view should be in the Location for duplicate drop down list. Select OK.

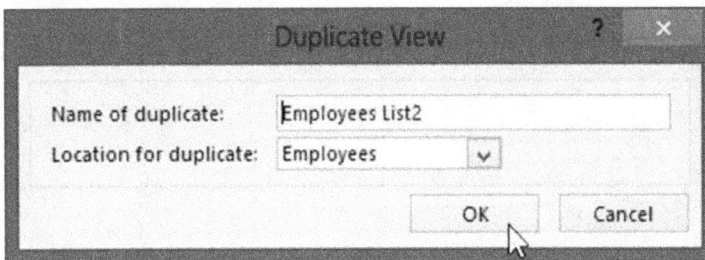

Adding a Popup View

Popup views help you get to information in one click.

To add a popup view, use the following procedure.

Step 1: Open the app in Access.

Step 2: Select the table and view that you want to edit.

Step 3: Select the Settings icon just to the left and below the View name.

Step 4: Select Edit.

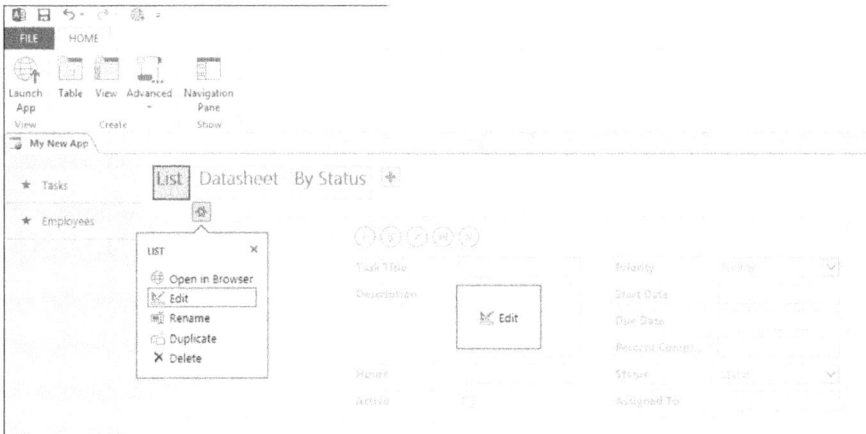

Step 5: Access opens the table for editing. Select the control to which you want to add the popup view. In this example, we will use the Assigned To control, because it is an AutoComplete control.

Step 6: Select the Data button next to the control.

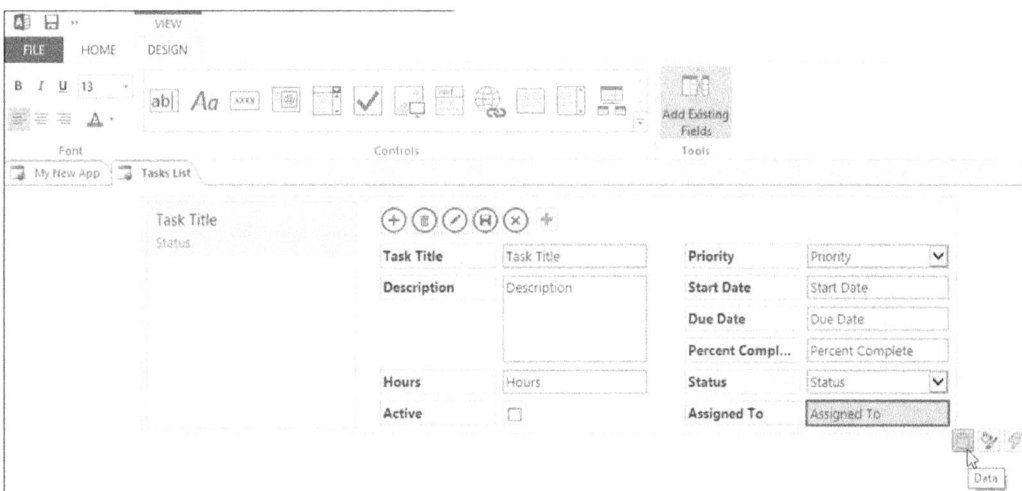

Step 7: If the control is bound to a row source (a table or query), you will see a Popup view drop-down near the bottom of the Data properties. The dropdown

shows views that have the same row source as the control. Select the view you want to appear when you click the popup button.

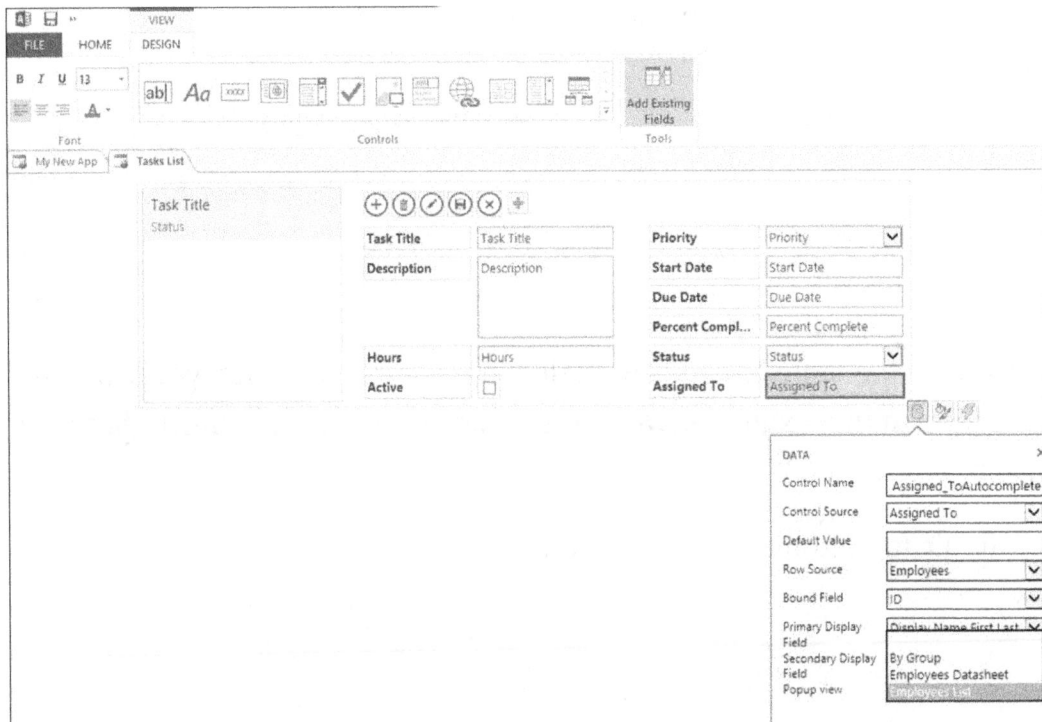

Step 8: Click the X at the top right of the Data dialog box to close it.

Step 9: Select Save to save your changes.

Step 10: Select the Home tab from the Ribbon.

Step 11q: Select Launch App.

To use the popup form, use the following procedure.

Step 1: Open your SharePoint or Office 365 account site.

Step 2: Open the app you just launched.

Step 3: In the Tasks table, select the employee name.

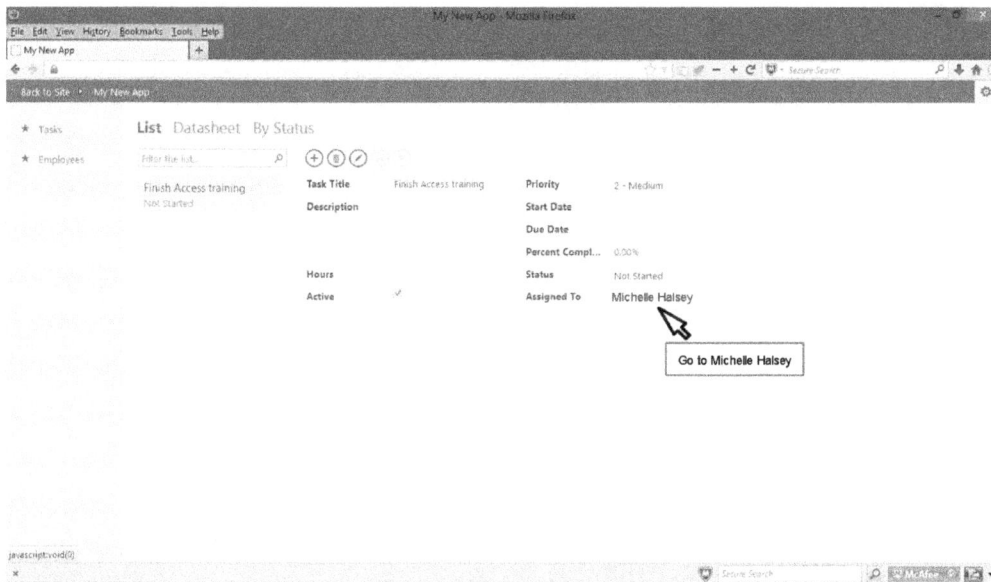

Step 4: The Employee popup form appears.

Employees

✕

Employee Num... 001   First Name Michelle

Last Name Halsey   Company

Email   Job Title

Work Phone   Home Phone

Mobile Phone   Address 1

Address 2   City

State/Province   ZIP/Postal Code

Country/Region   Office Location

Department   Date of Hire

Date of Birth   Emergency Con...

Emergency Con...   Emergency Con...

Web Page   Notes

Group Full-time

Tasks

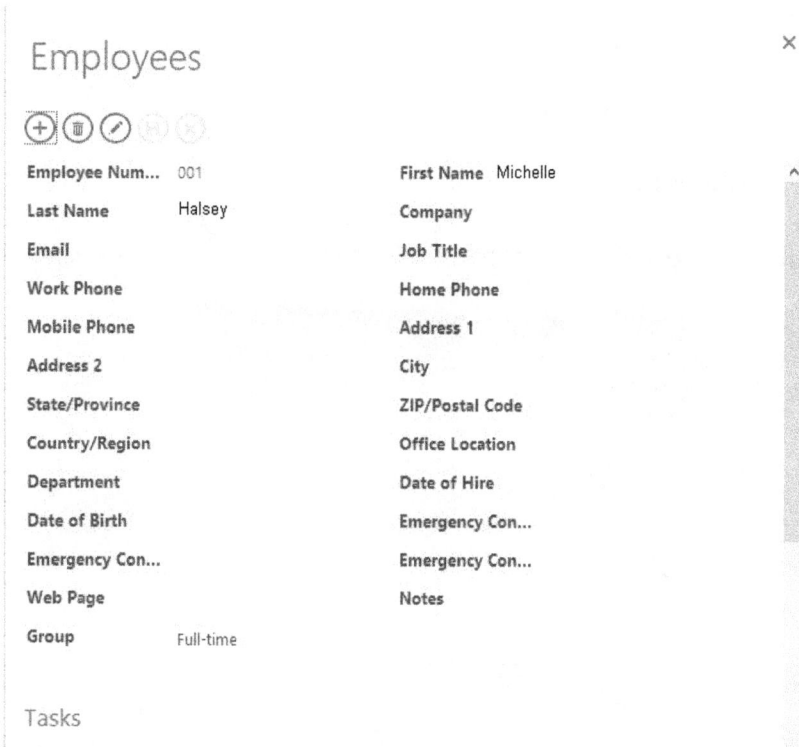

Deleting a View

You can delete a view if you no longer need it.

To edit a view, use the following procedure.

Step 1: Open the app in Access.

Step 2: Select the table and view that you want to delete.

Step 3: Select the Settings icon just to the left and below the View name.

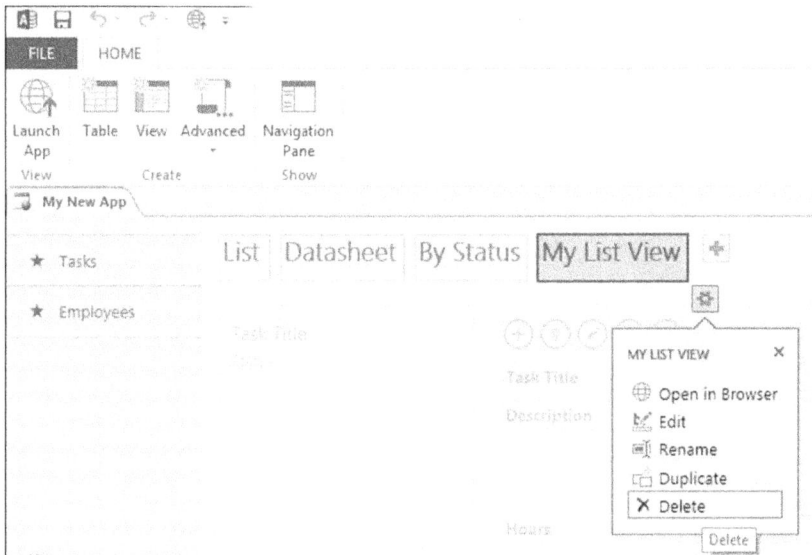

Step 4: Select Delete.

Step 5: In the confirmation dialog box, select Yes to continue.

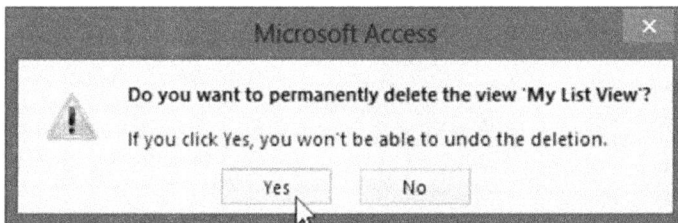

Step 6: Access permanently deletes the view.

Chapter 28: Customizing Apps

In this chapter, we will look at editing your tables to modify the action bar and moving, resizing, or deleting controls from your forms. We will also look at the Info tab on the Backstage View for information about your database. You can add unbound or bound controls, either from the Ribbon or from the Field List that is available when you are editing a table. You can also modify the control properties, which is especially useful if you want to change an unbound control to a bound one that updates the underlying table.

Opening a Table for Editing

In this chapter, we will do some basic editing for your app tables before getting into more advanced editing in the next chapter.

To open a table for editing, use the following procedure.

Step 1: If you are viewing the app in the browser, select the Settings icon on the top right corner of the screen and select Customize in Access.

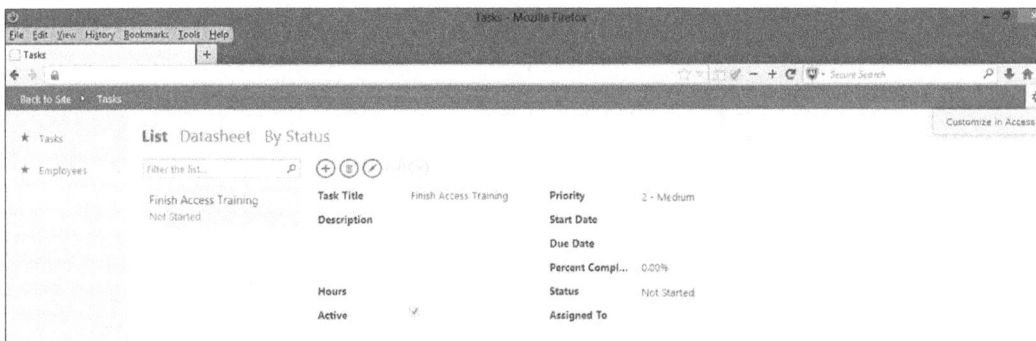

Step 2: Once the app opens in Access, select the table you want to modify.

Step 3: Select Edit Table.

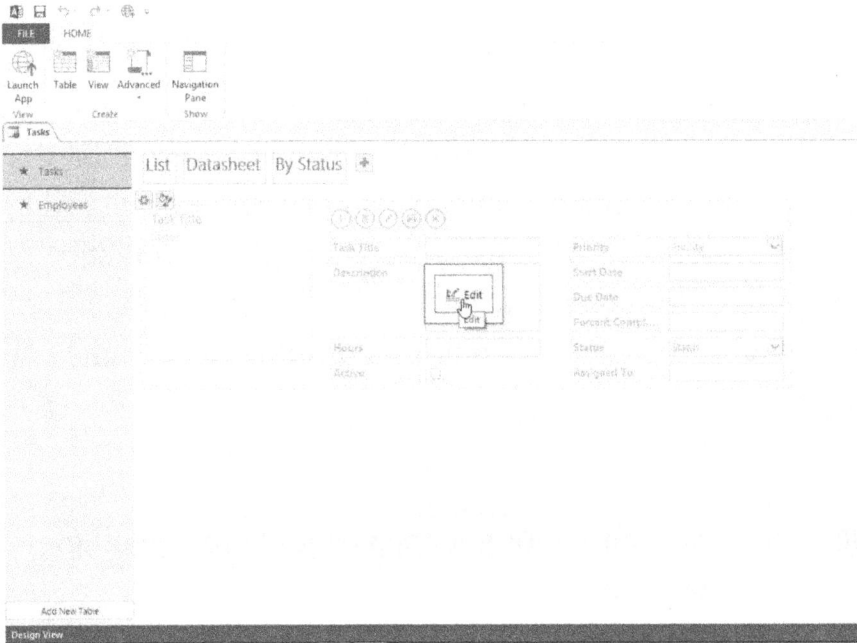

The table opens for editing. Each control on the table is now an active item that can be modified.

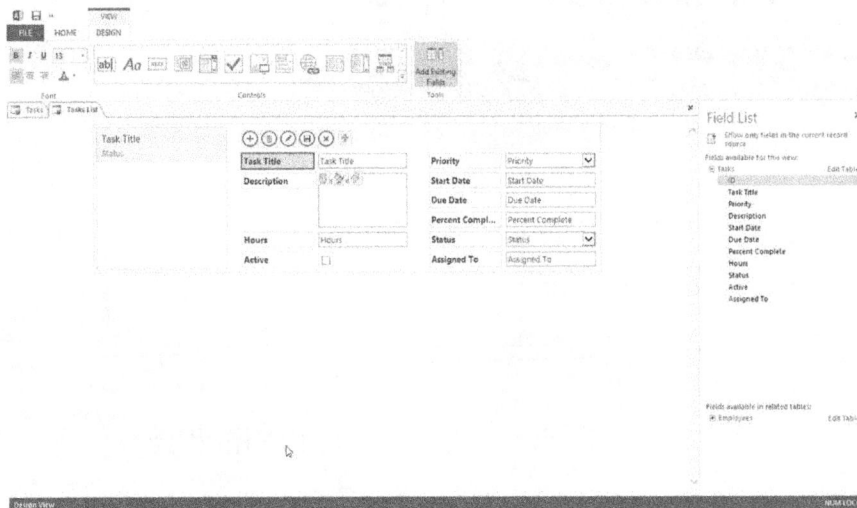

The Design tab on the View Ribbon.

Modifying the Action Bar

For now, we will simply look at changing the order of the actions on the Action bar.

To modify the action bar, use the following procedure.

Step 1: Open the table for editing as in the previous lesson.

Step 2: Click the Action Bar icon that you want to move.

Step 3: Drag it to the new location.

Step 4: Save the changes to the table.

Formatting Controls

The controls are the fields and labels on your table. You can format them to change the way they look or the visibility or tooltip.

To format controls, use the following procedure.

Step 1: Click on the control (field or label) that you want to format.

Step 2: To change the font size or add enhancements, make a selection from the Design tab on the View Ribbon.

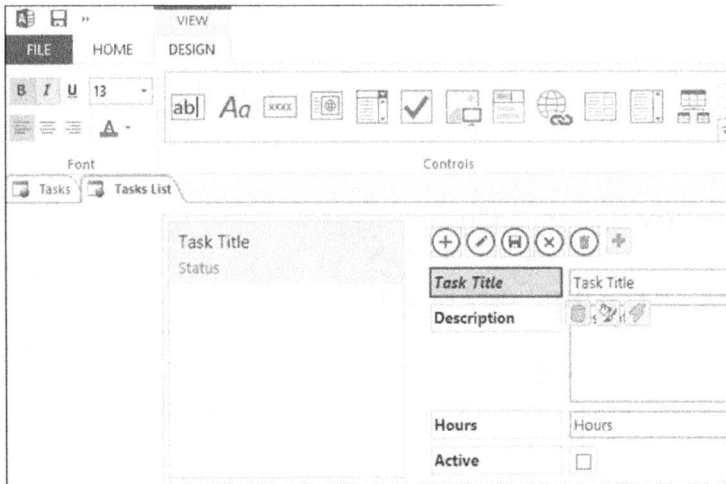

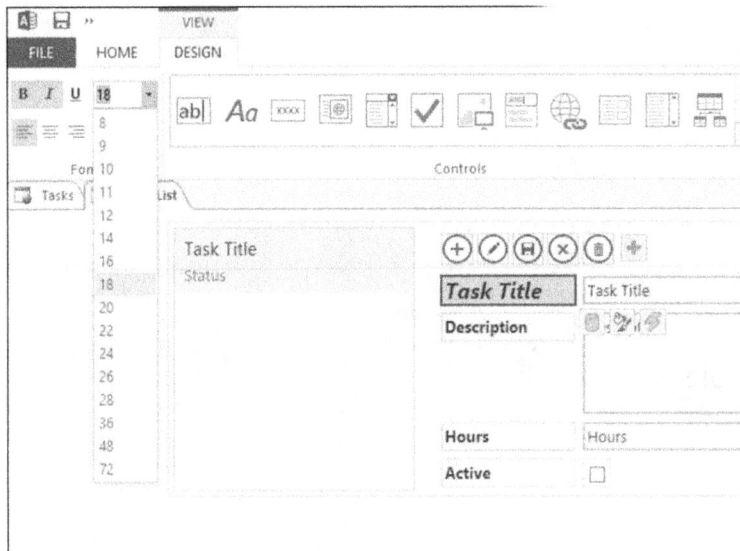

Step 3: To change the alignment within the area of the control, select the left, center, or right icons from the Design tab on the View Ribbon.

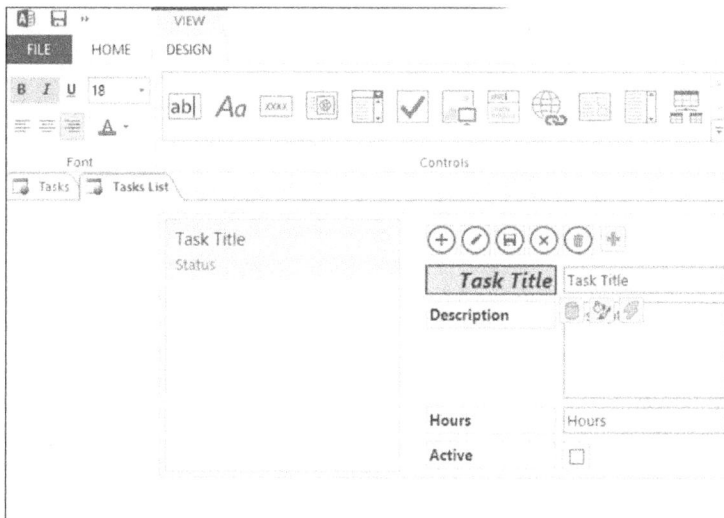

Step 4: To change the color of the text, select the Color tool from the Design tab on the View Ribbon.

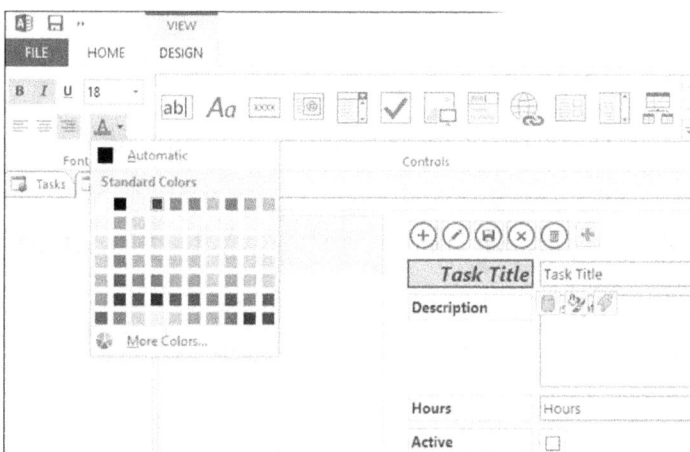

To change the Formatting Properties, use the following procedure.

Step 1: Click the control (field or label) that you want to format.

Step 2: Select the Formatting icon that pops up to the right of the control.

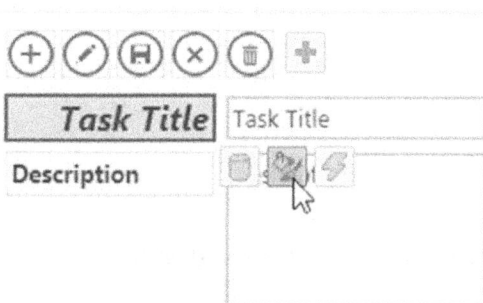

Step 3: In the Formatting dialog box for a label:

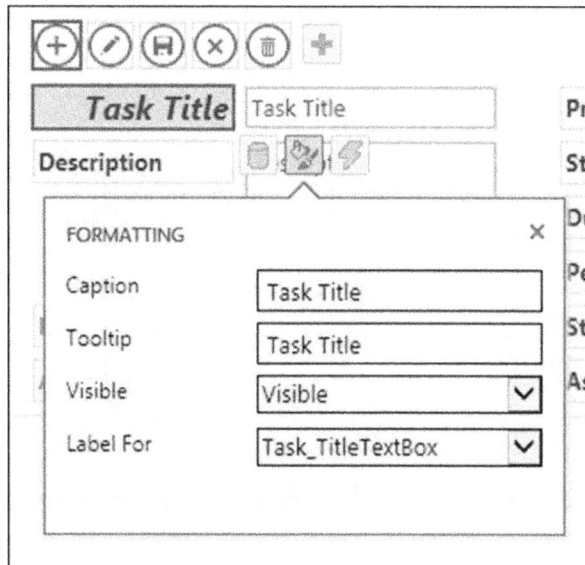

- You can edit the Caption to display different text on the control.
- You can also edit the Tooltip to display different text when the mouse hovers over the control
- Select an option from the Visible drop-down list to control whether the field is hidden or visible.
- The Label For drop down list presents a list of fields. The field selected should be the field near the selected label.

Step 4: In the Formatting dialog box for a field:

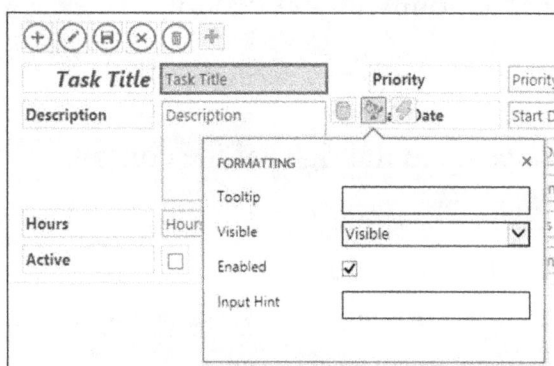

- You can edit the Tooltip to display different text when the mouse hovers over the control

- Select an option from the Visible drop-down list to control whether the field is hidden or visible.
- Check the Enabled box to enable the field.
- Enter text in the Input Hint field to display text in the field before the user has entered anything. This option is not available for some types of controls.

Step 5: Click the X in the top right of the Formatting dialog to close it.

Moving, Resizing, or Deleting Controls

In addition to formatting, you can move, resize, or delete the controls on your table.

To move controls, use the following procedure.

Step 1: Click on the control (field or label) that you want to move.

Step 2: Drag it to the new location on the table.

To resize a control, use the following procedure.

Step 1: Click on the control (field or label) that you want to resize.

Step 2: Drag the side of the control bigger or smaller as desired. Click again when the control is your desired size.

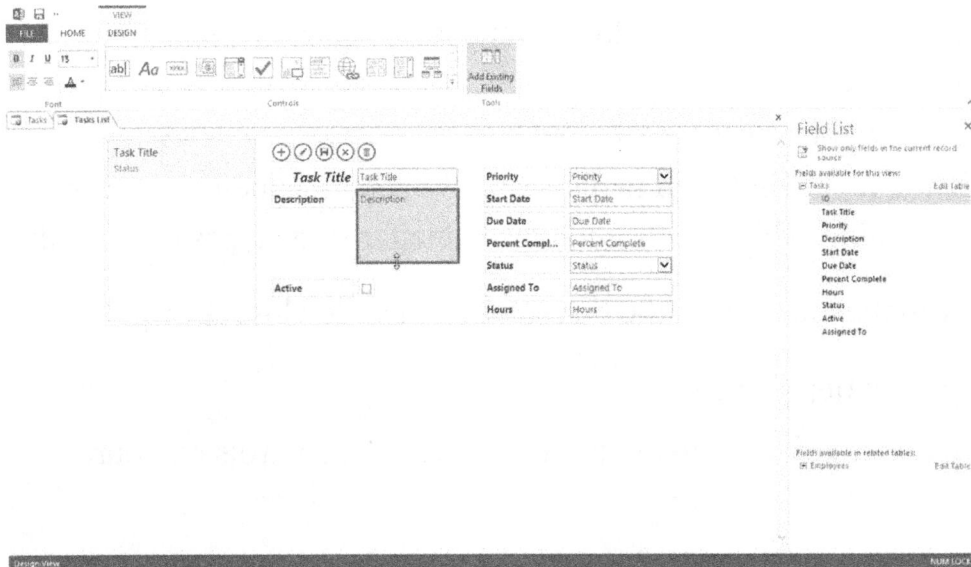

To delete a control, Use the following procedure.

Step 1: Click the control (field or label) that you want to delete.

Step 2: Press the Delete or the Backspace key.

Step 3: Rearrange the remaining fields as desired.

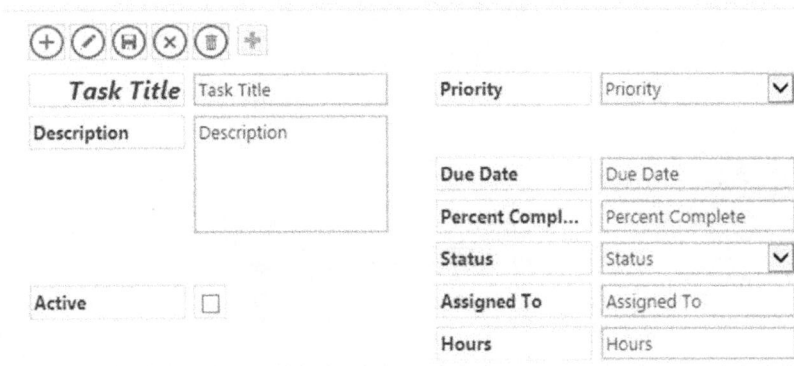

Adding Controls from the Ribbon

You can a variety of controls right from the Ribbon.

To add a control from the Ribbon, use the following procedure.

Step 1: Open the table you want to modify for editing.

Step 2: Select a control from the Design tab on the View Ribbon.

Step 3: Access places the control on your view. You may need to move it.

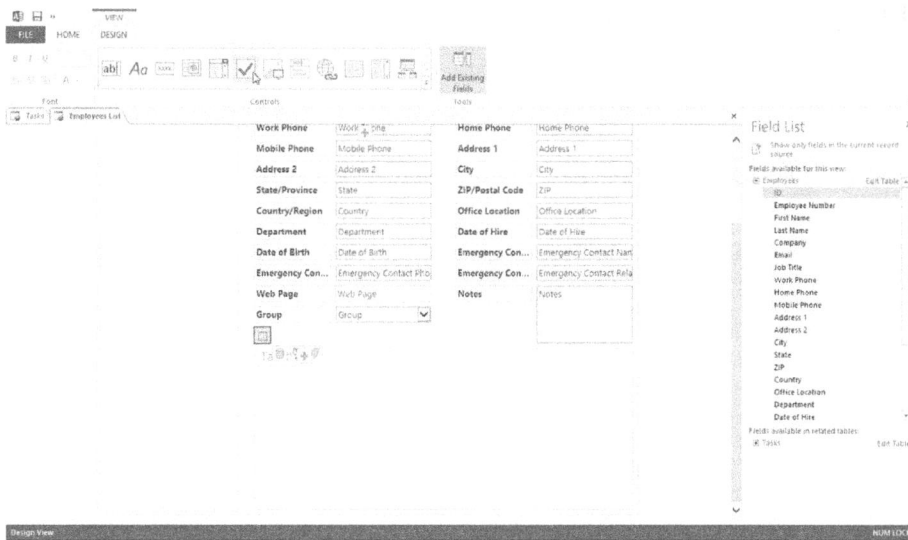

Adding Controls from the Field List

We will learn about changing the control properties in the next lesson in case you want to change the controls you just added to bound controls. But the easiest way to get a bound control is to add it from the Field List.

To add a control from the Field List, use the following procedure.

Step 1: Open the table you want to modify for editing.

Step 2: Select a Field from the Field List and drag it to the view.

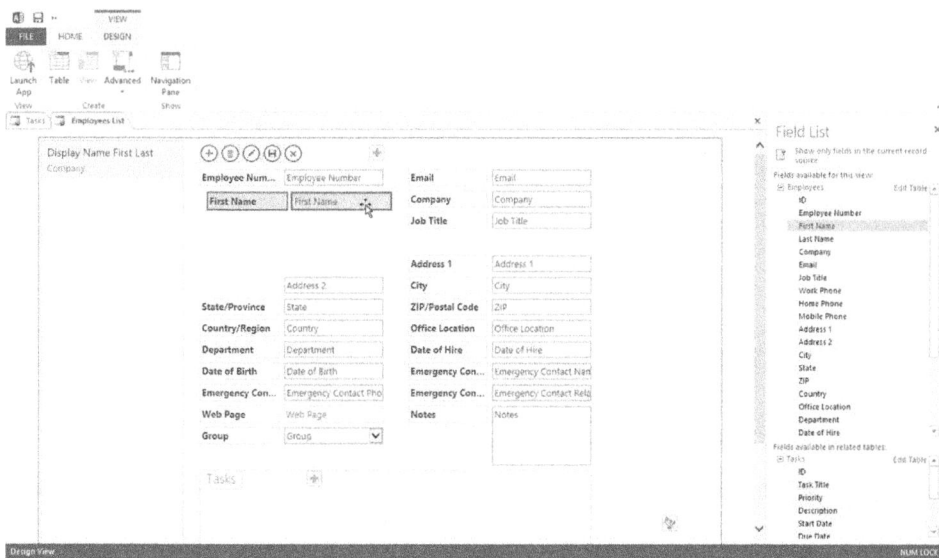

Changing Control Properties

In this lesson, we will look at changing the Data properties.

To change control properties, use the following procedure.

Step 1: Click on the control (field or label) that you want to change.

Step 2: Select the Data icon that pops up to the right of the control.

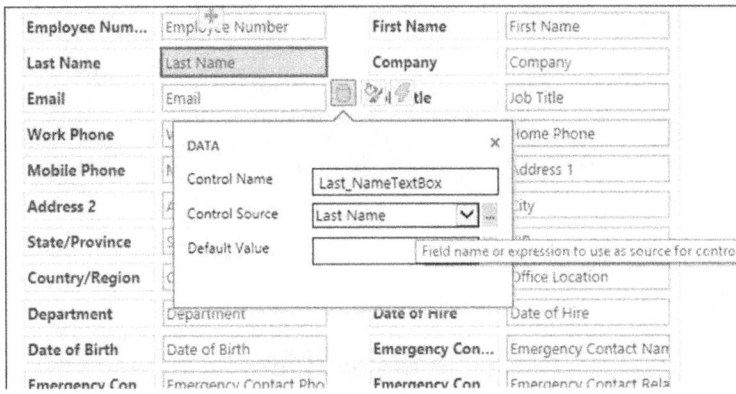

Step 3: In the Data dialog box, you can change the Control name that is used by expressions and macros. Be careful changing this, because it may be referenced elsewhere in your app.

Step 4: To change the binding or create one for an unbound control, select a new field from the Control Source drop down list.

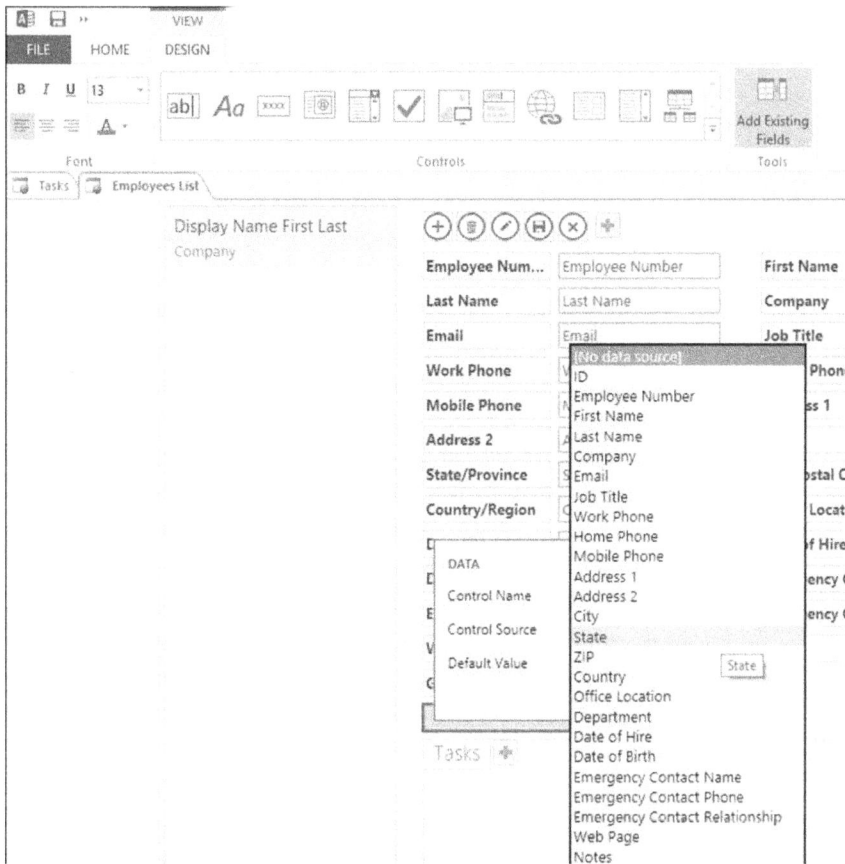

Step 5: To include a default value (which the user can change, if needed), enter the Default Value.

Step 6: Click the X in the top right of the Data dialog to close it.

Step 7: Save the changes to your table.

Chapter 29: Customizing Your App Views with Macros

Interface macros can perform actions like opening another view or applying a filter. An embedded user interface macro attaches directly to objects in your app like buttons, combo boxes, or the Action Bar button. Standalone user interface macros can be reused to avoid duplicating code. You can run a standalone macro from within an embedded macro. You will learn how to create these macros in this chapter. We will also learn how to save your app as a package.

About User Interface Macros

This lesson will introduce user interface macros.

Embedded UI macros run when specific events occur in a view, such as clicking a button, selecting an item in a combo box, or loading a view. The macros become part of the view or control they are embedded in.

Here are the events you can attach a UI macro to in a control or view:

EVENT TYPE	WHEN IT OCCURS
After Update	Occurs after you type data into a control or select an item in a list control.
On Click	Occurs when a control is selected.
On Current	Occurs when the user moves to a different record in the view.
On Load	Occurs when a view is opened.

And here are the events each control supports:

CONTROL OR OBJECT TYPE	SUPPORTED EVENTS
Action Bar Button	On Click
AutoComplete	After Update, On Click
Button	On Click
Check Box	After Update
Combo Box	After Update

Hyperlink	After Update, On Click
Image	On Click
Label	On Click
Multiline Textbox	After Update, On Click
Text Box	After Update, On Click
View	On Current, On Load

Standalone macros can help you avoid duplicating code. You will create the standalone macro independent of any object. Once you have created a stand-alone macro, it is available in the Action Catalog to call from other macros.

Creating an Embedded Macro

Embedded macros are attached to objects in your app, like action buttons, or controls on your view like buttons or combo boxes.

To create an embedded user interface macro, use the following procedure.

Step 1: Open the view you want to modify.

Step 2: Select the control that will trigger the macro (such as a button).

Step 3: Select the Actions icon to the right and below the selected control.

Step 4: In the Actions dialog box, select the option for opening the Macro Builder. Depending on the type of control, you may see After Update or On Click (or even both).

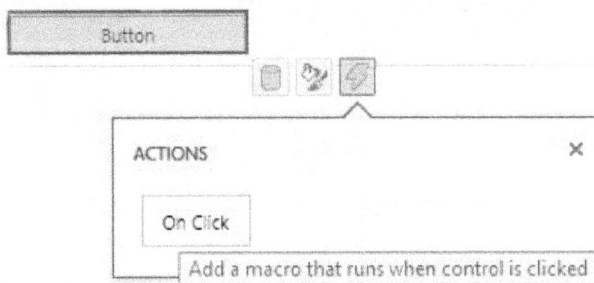

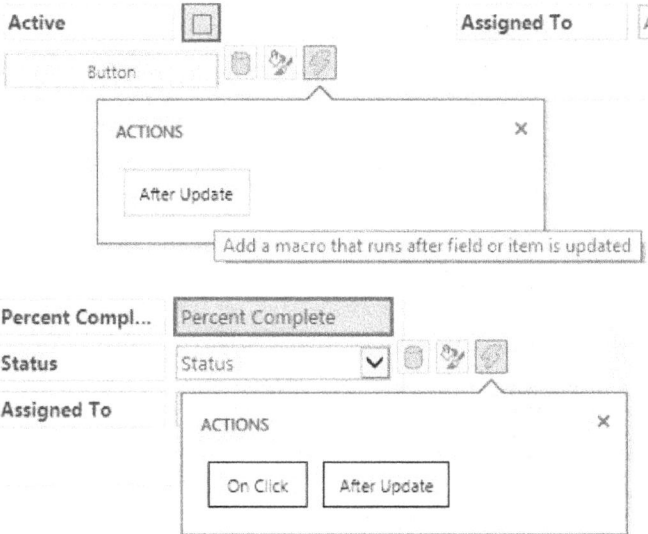

Step 5: In the Macro Design view, select from the drop down list the first action that you want the control on the selected view to perform. You can also add macro actions from the Action Catalog on the right. Just double-click on the action you want to include. Notice that if you click once on an action in the Action Catalogue, Access displays a definition at the bottom of the Action catalog pane.

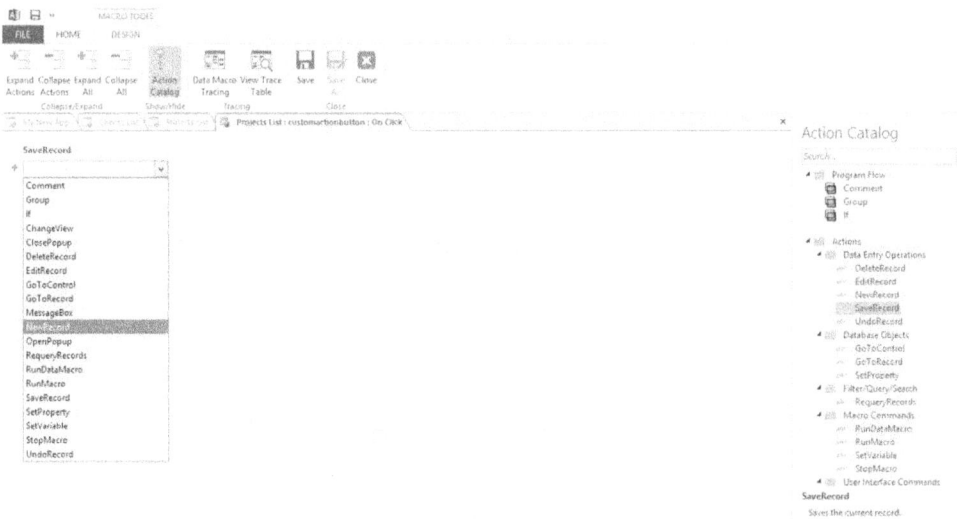

Step 6: If you select actions that require the arguments, fill in the argument information.

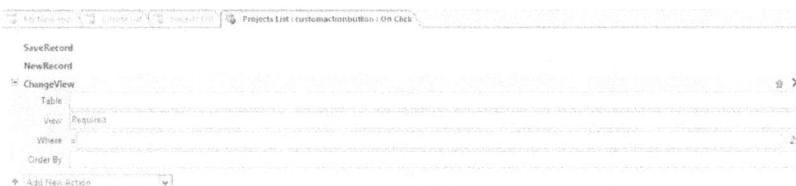

Step 7: If you add multiple actions, you can use the arrow keys on the right to rearrange them. You can also click the X to delete an action you no longer need.

Step 8: Save your changes and close the macro design view.

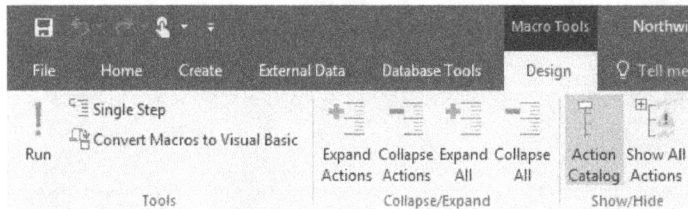

Step 9: In the Actions dialog box, the button for the type of Macro you created (On Click or After Update) is now green to indicate that you have added actions.

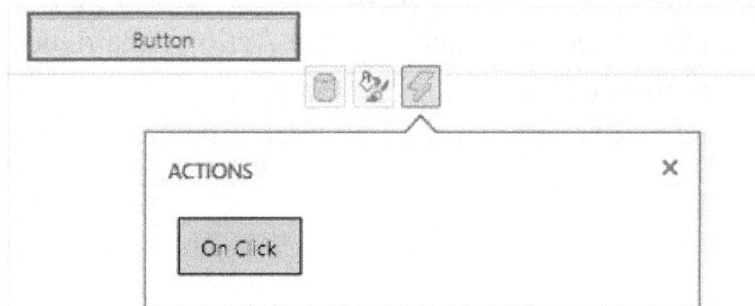

Step 10: Also make sure to save your database and launch the app to upload your changes to the browser.

Creating a Standalone Macro

A standalone macro is not tied to an object in your app, so it can be reused.

To create a standalone macro, use the following procedure.

Step 1: Select the Home tab from the Ribbon.

Step 2: Select Advanced.

Step 3: Select Macro.

Step 4: In the Macro Design view, select the first action you want the standalone macro to perform from the drop-down list. You can also add macro actions from the Action Catalog on the right. Just double-click on the action you want to include. Notice that if you click once on an action in the Action Catalogue, Access displays a definition at the bottom of the Action catalog pane.

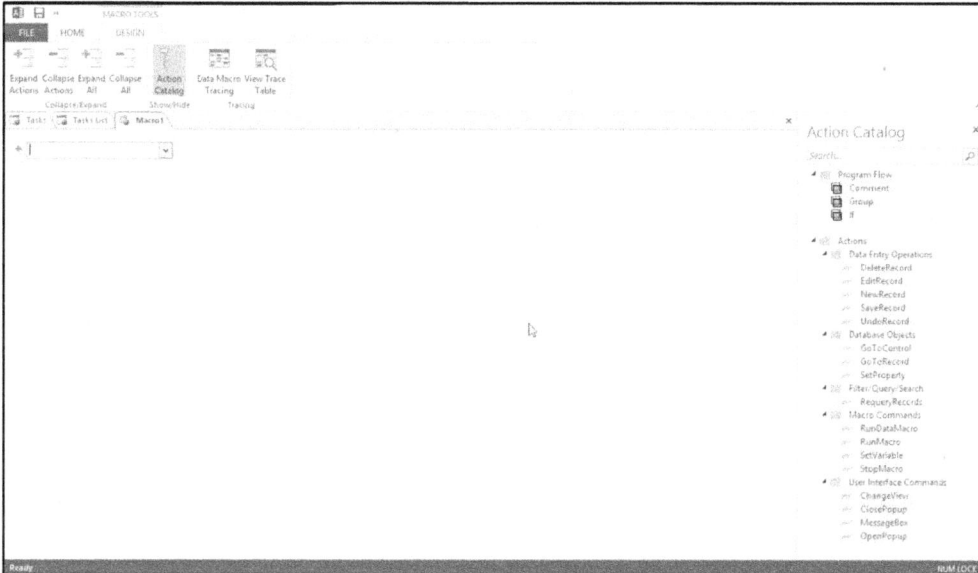

Step 5: If you select actions that require the arguments, fill in the argument information.

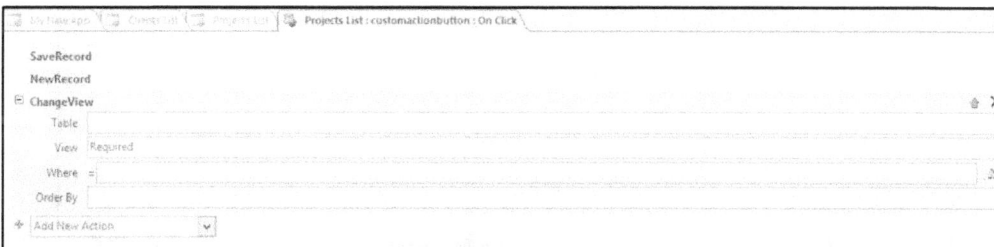

Step 6: If you add multiple actions, you can use the arrow keys on the right to rearrange them. You can also click the X to delete an action you no longer need.

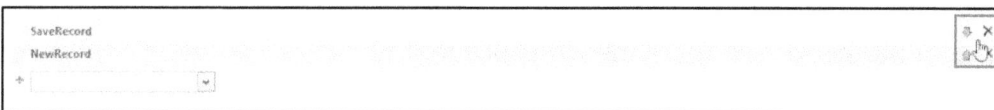

Step 7: Save your changes and close the macro design view.

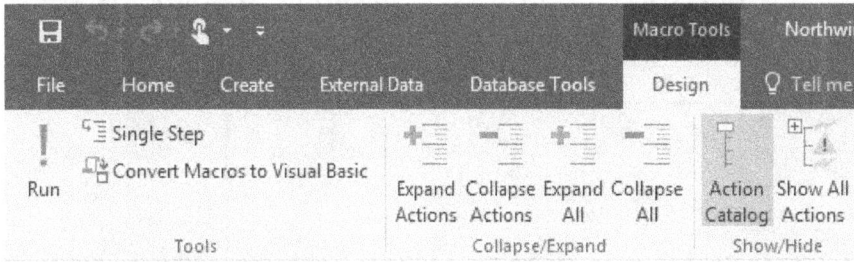

Review the new category and macro now available in the Action Catalog for an embedded macro.

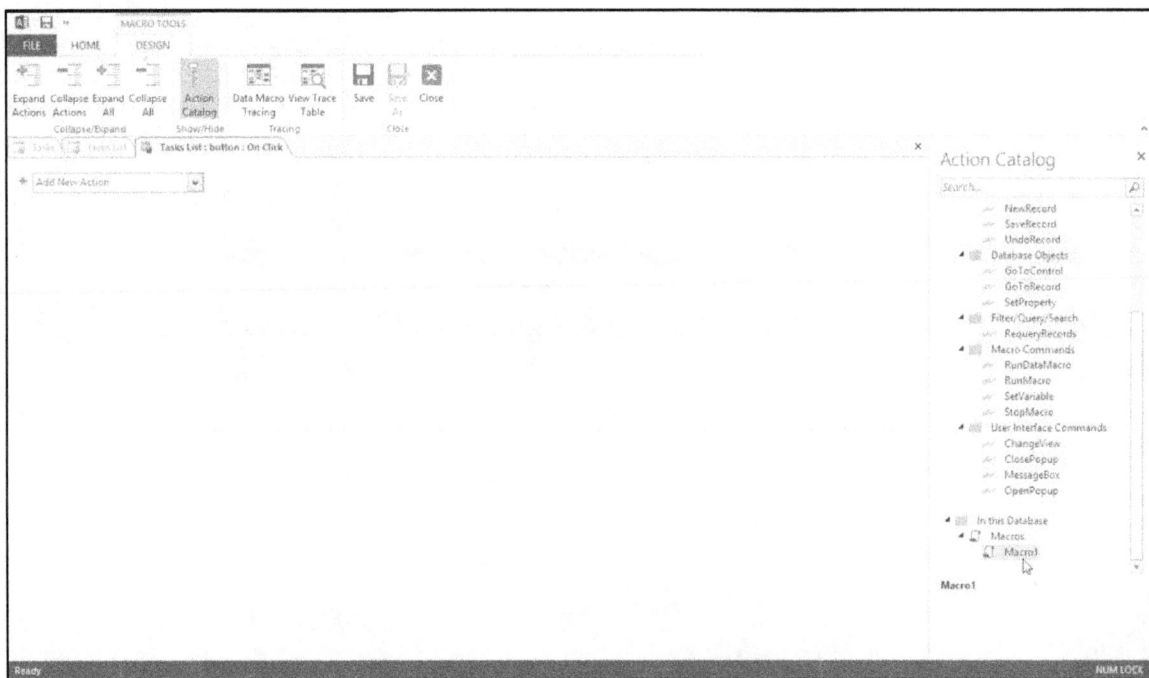

Saving Your App as a Package

Saving your app as a package allows you to back it up, move it or deploy it.

To save the app as a package, use the following procedure.

Step 1: Select the File tab from the Ribbon.

Step 2: Select Save As.

Step 3: Select Save As Package.

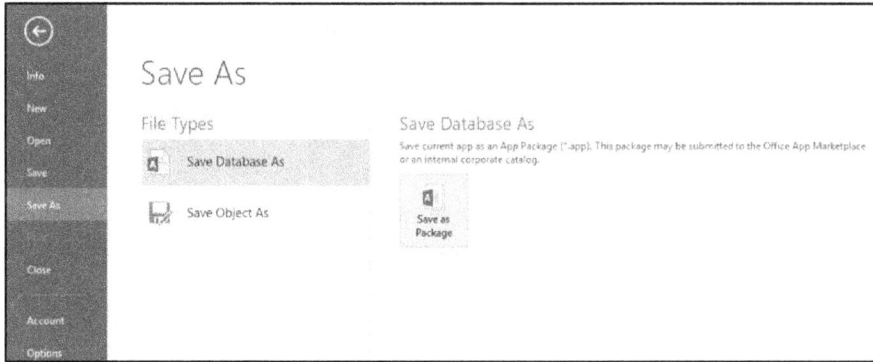

Step 4: Enter a title for the package.

Step 5: Check the Include Data in the Package box if you want to save the data along with the structure.

Step 6: Select OK.

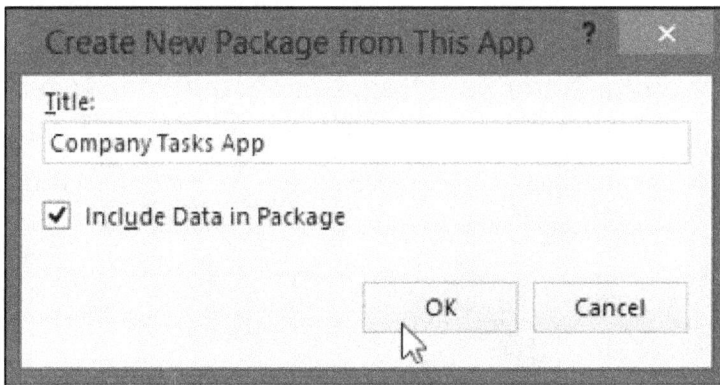

Step 7: Navigate to the location where you want to save the app package and select OK.

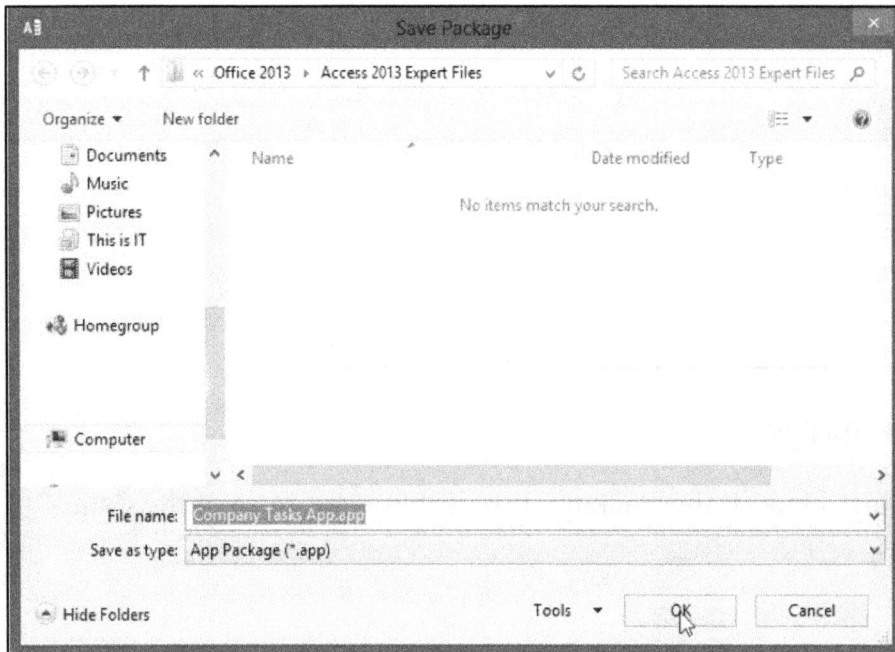

Once you have saved the app package, it can be loaded to any SharePoint Online site via the App catalog. Owner or Administrator permissions are required to make an app available on the catalogue.

Chapter 30 – Microsoft Access 2016 New Features

dBase Support Available

Microsoft Access 2016 has added support for DBASE (.dbf) support which allows you to import or link to data stored in a dBase file or export data to a dBase file.

Import & Link dBase File

Step 1: Click the External Data tab of the Ribbon.

Step 2: Click the arrow next to New Data Source to expand the data source options.

Step3: Choose From Database and then from dBase File in the menu.

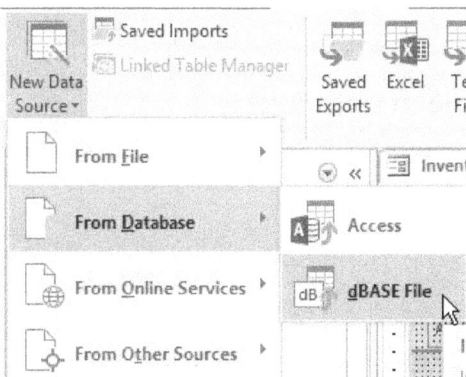

Step 4: Browse to the dBase file and choose how you want to import the file.

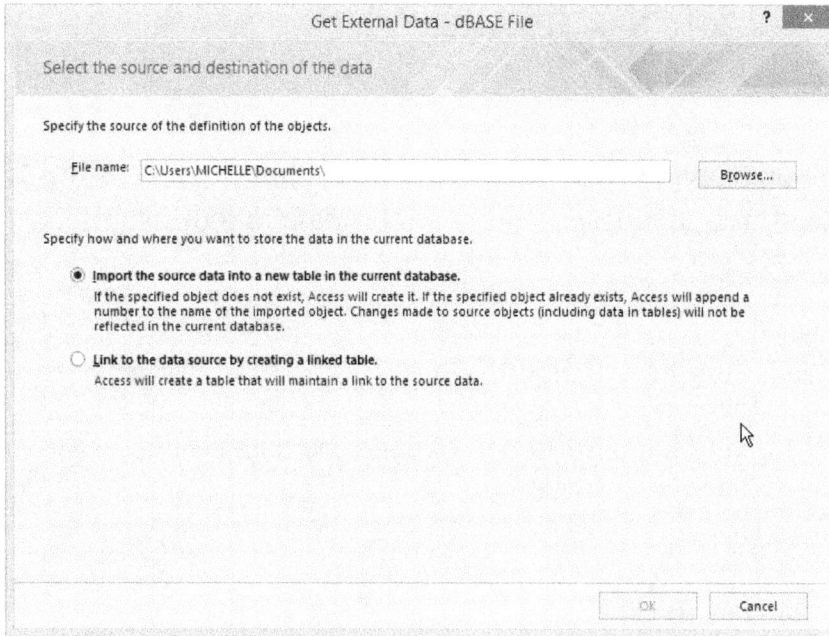

Step 5: Click Ok to import the file.

Export dBase File

Step 1: Click the External Data tab of the Ribbon.

Step 2: Click the arrow next to More in the Export menu option.

Step 3: Choose dBase file from the drop-down menu and follow the prompts to export the file.

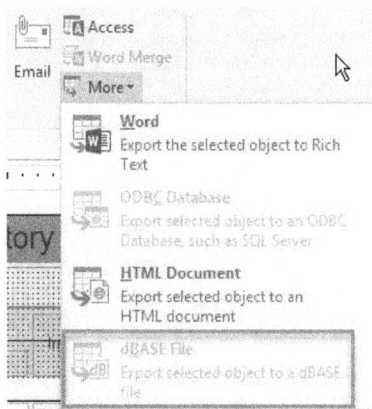

Export linked data source information to Excel

Access now allows you to export a list of the data sources you are using to an Excel spreadsheet. This can be helpful if you are working on a complex database application that includes links to many data sources.

Step 1: Click the External Data tab of the Ribbon.

Step 2: Click the Linked Table Manager option in the Import & Link menu.

Step 3: Select the linked data sources you want to export and click the Export to Excel button.

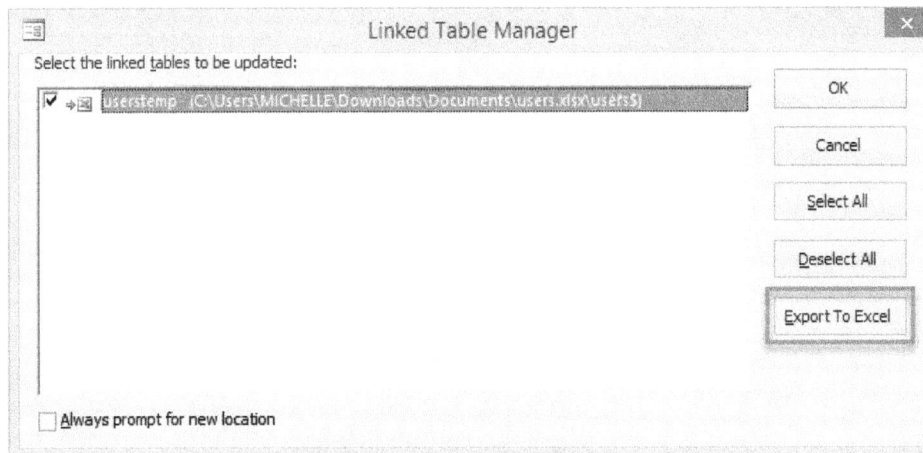

Step 4: You will be prompted to select a location to save the file. Enter a name for the file and click Save.

Large Number Data Type Added

The large number (8 bytes) data type is compatible with the SQL_BIGINT data type in an OBDC and is used to store a numeric value. This data type can be added as a field as an Access table and is used to calculate large numbers. The large number data type can handle a range of -263 to +263. The Number data type in contrast can only handle a range of -231 to +231.

This data type allows an Access database to work more efficiently with linked or imported data. Before you can use the Large Number data type when linking or importing from an external source, you will need to enable the "Support Bigint Data Type for Linked/Imported Tables" option.

Step 1: Click the File tab on the Ribbon.

Step 2: Click Options.

Step 3: Click the Current Databases menu and scroll down the page to the Data Type Support Options section.

Step 4: Place a checkmark in the box next to "Support Large Number (Bigint) Data Type for Linked/Imported Tables" and click Ok.

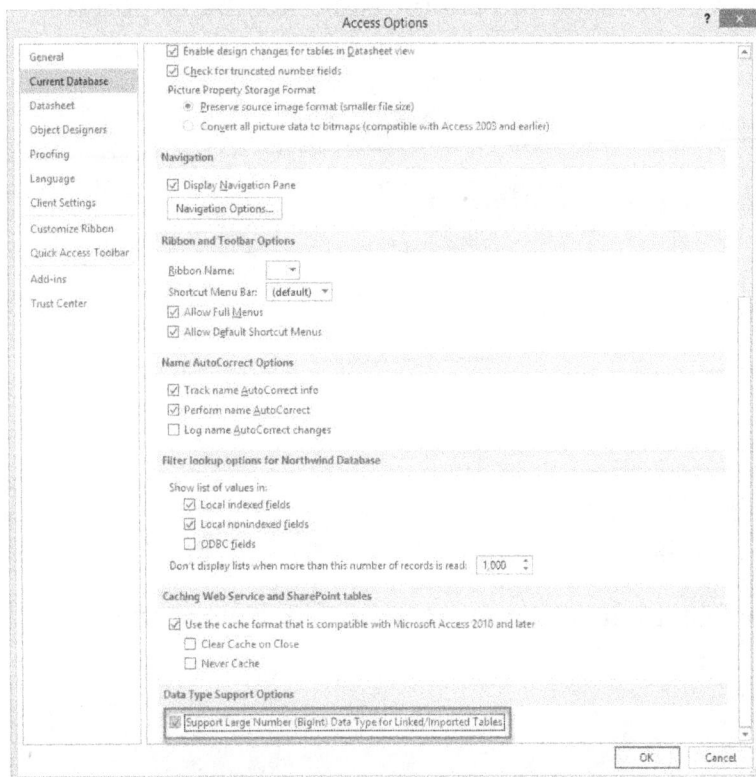

When you check this option, you may receive a message box warning you of the loss of backwards compatibility for the database.

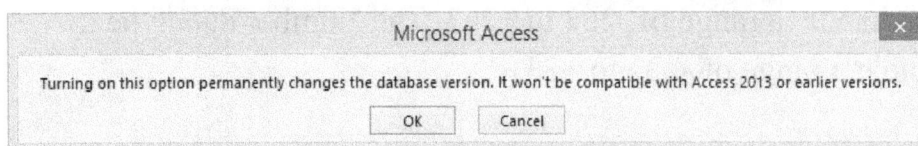

If you add the large number data type to a database, this will eliminate compatibility with older versions of the database, so the decision to use this data type should keep this in mind. Enabling this data type does not automatically change the data type of existing tables in your database. If you had connected to a database with a large number in the past, then Access would have converted this field to a short text field. To update the field, you will need to refresh the linked table or update the imported table.

To refresh the linked table, perform the following:

Step 1: Click the External Data tab on the Ribbon.

Step 2: Click the Linked Table Manager.

Step 3: Select the tables you want to refresh and then click Ok. This converts the Short Text data type to the Large Number data type.

To update the imported table, perform the following:

Step 1: Open the table in Design View.

Step 2: Choose Large Number from the list of data types in the Data Type column.

Step 3: Click Save to save the changes.

New feedback Option

Like with the other applications in the Microsoft Suite, there is a new feedback option. To provide feedback, perform the following:

Step 1: Click the File tab on the ribbon.

Step 2: Click Feedback on the left navigation menu.

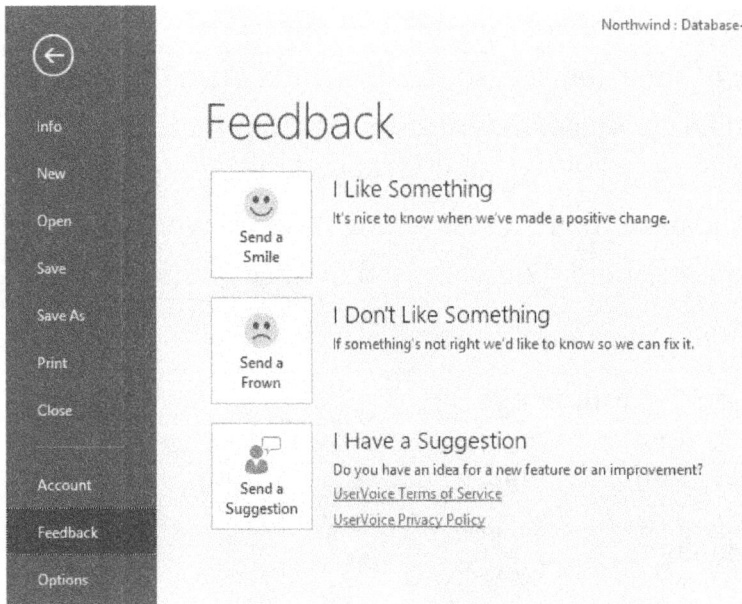

Step 3: Select one of the three feedback options.

Step 4: Fill in the feedback form and click submit.

New Label Name Property for Form Controls

Microsoft Access 2016 has added the property Label Name to form controls, so a label control can be associated with another control. You can now type the name of the label control rather than cutting and pasting the control to create the association between the controls. This is an advantaged for accessibility because assistive technologies can now detect the association.

276

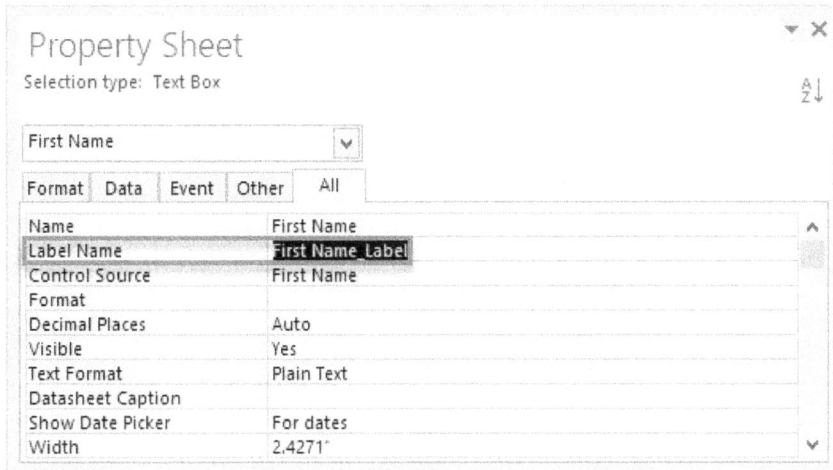

Property Sheet

Selection type: Text Box

| First Name | ⌄ |
| Format | Data | Event | Other | All |

Name	First Name
Label Name	First Name_Label
Control Source	First Name
Format	
Decimal Places	Auto
Visible	Yes
Text Format	Plain Text
Datasheet Caption	
Show Date Picker	For dates
Width	2.4271"

New Shortcut for Editing a New Value List Item

A new shortcut has been added making it easier to open the Edit List Items dialog box when you are working on a value list combo box in an Access Form. Two criteria are needed: the combo box uses a value list as the data source and the "Allow Value List Edits" property is set to Yes. While in Form view with the focus on the combo box, press CTRL + E to open the Edit List Items dialog.

Property Sheet Sorting

If you are having difficulty finding a specific property on the Property sheet for a form or report, you can now sort the properties. The default in Access shows the properties unsorted, in the order you would typically see the properties. Perform the following to sort the field.

Step 1: Open the form or report in Design View.

Step 2: Right-click the field and select Properties from the drop-down menu.

Step 3: Click the Sort toggle button on the top right of the Property Sheet to sort the properties alphabetically.

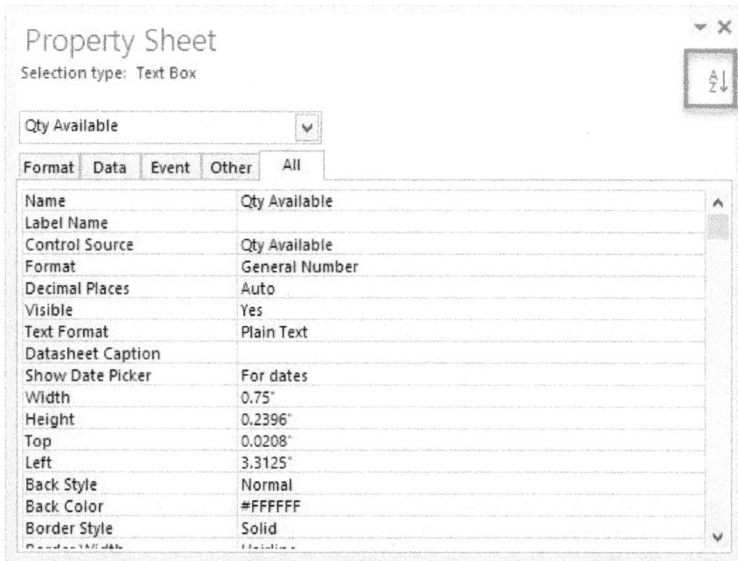

Clicking Sort a second time will return the properties to their original order.

Tell Me

Instead of searching the online help or in Access, you can use the Tell Me feature to look for the solution you need. Click in the "tell me what you want to do" area and type your request.

Chapter 31 – Microsoft Access 2016 Shortcuts

Copy, Move, or Delete Text

To do this	Press
Delete selection or character to the left of insertion point	Backspace
Copy the selected control to the Clipboard or text in cell	CTRL + C
Delete all characters to the right of the insertion point	CTRL + Delete
Paste contents of Clipboard at insertion point	CTRL + V
Cut the selection and copy it to the Clipboard	CTRL + X
Delete selection or character to right of insertion point	Delete

Edit Text or Data

To do this	Press
Move insertion point to end of field, in multi-line fields	CTRL + End
Move insertion point to beginning of field, in multi-line fields	CTRL + Home
Move insertion point 1 word to the left	CTRL + Left Arrow
Move insertion point 1 word to right	CTRL + Right Arrow
Move insertion point to end of field, in single-line fields; or move it to end of the line in multi-line fields	End

To do this	Press
Move insertion point to beginning of field, in single-line fields; or move it to beginning of line in multi-line fields	Home
Move insertion point 1 character to left	Left Arrow
Move insertion point 1 character to right	Right Arrow

Edit in a Text Box

To do this	Press
Move 1 word to left or right	CTRL + Left Arrow or CTRL + Right Arrow
Change selection by 1 word to left	CTRL + SHIFT + Left Arrow
Change selection by 1 word to right	CTRL + SHIFT + Right Arrow
Move to end of entry	End
Move to beginning of entry	Home
Select from insertion point to end of text entry	SHIFT + End
Select from insertion point to beginning of text entry	SHIFT + Home
Change selection by 1 character to left	SHIFT + Left Arrow
Change selection by 1 character to right	SHIFT + Right Arrow

To do this	Press
Move 1 character to left or right	Left Arrow or Right Arrow

Edit & Navigate Object List

To do this	Press
Move to last object	End
Rename selected object, when object is closed	F2
Move down 1 window	Page Down
Move up 1 window	Page Up
Move down 1 line	Down Arrow
Move up 1 line	Up Arrow

Edit Using Controls in Form and Report Design View

To do this	Press
Paste contents of Clipboard in upper-left corner of selected section	CTRL + V
Cut selected control and copy to Clipboard	CTRL + X
Increase height of selected control NOTE: If used with a control that is in a layout, the entire row of the layout is resized.	SHIFT + Down Arrow
Reduce width of selected control	SHIFT + Left Arrow

To do this	Press
NOTE: If used with a control that is in a layout, the entire column of the layout is resized.	
Increase width of selected control	SHIFT + Right Arrow
NOTE: If used with a control that is in a layout, the entire column of the layout is resized.	
Reduce height of selected control	SHIFT + Up Arrow
NOTE: If used with a control that is in a layout, the entire row of the layout is resized.	
Move selected control down (except a control that is part of a layout)	Down Arrow or CTRL + Down Arrow to move in smaller increments
Move selected control to left (except a control that is part of a layout)	Left Arrow or CTRL + Left Arrow to move in smaller increments
Move selected control to right (except a control that is part of a layout)	Right Arrow or CTRL + Right Arrow to move in smaller increments
Move selected control up (except a control that is part of a layout)	Up Arrow or CTRL + Up Arrow to move in smaller increments

Enter Data in a Datasheet or Form

To do this	Press
Insert default value for a field	CTRL + ALT + Spacebar
Insert value from same field in previous record	CTRL + ' (Apostrophe)
Insert new line in Short Text or Long Text field	CTRL + ENTER
Delete the current record in a datasheet	CTRL + - (Minus Sign)
Add new record	CTRL + + (Plus Sign)
Insert current date	CTRL + ; (Semicolon)
Insert current time	CTRL + SHIFT + : (Colon)
Save changes to current record	SHIFT + ENTER
Switch between values in check box or option button	Spacebar

Extend a Selection

To do this	Press
Cancel Extend mode	Esc
Turn on Extend mode NOTE: In Datasheet view, Extended Selection appears in the lower-right corner of the window	F8 (repeatedly pressing F8 extends the selection to the word, the field, the record, and all records)

To do this	Press
Undo previous extension	SHIFT + F8
Extend a selection to adjacent fields in same row in Datasheet view	Left Arrow or Right Arrow
Extend a selection to adjacent rows in Datasheet view	Up Arrow or Down Arrow

Find & Replace Text and Data

To do this	Press
Open Find tab in Find and Replace dialog box (Datasheet view and Form view only)	CTRL + F
Open Replace tab in Find and Replace dialog box (Datasheet view and Form view only)	CTRL + H
Find next occurrence of text specified in Find and Replace dialog box when dialog box is closed (Datasheet view and Form view only)	SHIFT + F4

Get Help

To do this	Press
Go back to Access Help Home	ALT + Home
Move back to previous Help topic (Back button)	ALT + Left Arrow or Backspace
Move forward to next Help topic (Forward button)	ALT + Right Arrow
Print current Help topic	CTRL + P (if the cursor is not in

To do this	Press
	the current Help topic, press F6, and then press CTRL + P)
Perform action for selected item	ENTER
Expand or collapse selected item in Access Help topics list	ENTER
Perform action for selected Show All, Hide All, hidden text, or hyperlink	ENTER
In Table of Contents in tree view, expand or collapse selected item	ENTER
Stop last action (Stop button)	Esc
Open Help Window	F1
Refresh window (Refresh button)	F5
Switch among areas in Help window, such as toolbar and Search list	F6
Change connection state	F6, and then press ENTER to open the list of choices
Scroll larger amounts up or down, respectively, within currently displayed Help topic	Page Up, Page Down
Select previous item in Help window	SHIFT + Tab
Select previous hidden text or hyperlink	SHIFT + Tab

To do this	Press
Select next item in Help window	Tab
Select next hidden text or hyperlink, including Show All or Hide All at top of a topic	Tab
In a Table of Contents in tree view, select the next or previous item, respectively	Up Arrow or Down Arrow
Scroll small amounts up or down within currently displayed Help topic	Up Arrow, Down Arrow

Miscellaneous

To do this	Press
Exit Access	ALT + F4
Copy a screenshot of current window to the Clipboard	ALT + Print Screen
Display full set of commands on task pane menu	CTRL + Down Arrow
Invoke Builder	CTRL + F2
Display complete hyperlink address (URL) for a selected hyperlink	F2
Check spelling	F7
Copy a screenshot of entire screen to Clipboard	Print Screen
Open Zoom box to enter expressions and other text in small input areas	SHIFT + F2

Move Focus

To do this	Press
	Alt or F10
Select active tab of ribbon and activate Key Tips	(to move to a different tab, use Key Tips or the arrow keys)
Expand or collapse ribbon	CTRL + F1
Finish modifying a value in a control on the ribbon, and move focus back to document	ENTER
Move focus to a different pane of window	F6
Display shortcut menu for selected item	SHIFT + F10
Activate selected command or control on ribbon	Spacebar or ENTER
Open selected menu or gallery on ribbon	Spacebar or ENTER
Open selected list on ribbon, such as Font list	Down Arrow
Move down, up, left, or right, respectively, among items on ribbon	Down Arrow, the Up Arrow key, Left Arrow, or Right Arrow
Move between items in open menu or gallery	Tab
Move focus to commands on ribbon or move to next or previous command on ribbon	Tab or SHIFT + Tab

Navigate the Ribbons

To Do This	Key
Move between groups on a ribbon	CTRL + Right Arrow or CTRL + Left Arrow
Move to list of ribbon tabs	Alt
Open Create tab	ALT + C
If Activated command selected is a split button (a button that opens a menu of additional options), tab through the options. To select the current option, press Spacebar or Enter.	ALT + Down Arrow
Open File Page	ALT + F
Open Home tab	ALT + H
Open Fields tab	ALT + J, B
Open Table tab	ALT + J, T
Open Tell me box	ALT + Q, and then enter the search term
	SHIFT + F10
Open External Data tab	ALT + X or ALT + X,1
Open Add-ins tab, if present	ALT + X, 2
Open Database Tools tab	ALT + Y
To move in group currently selected	Down Arrow Key

To Do This	Key
To activate selected command when it is a button	Spacebar or ENTER
To select command if selected command is a gallery. Then, tab through items.	Spacebar or ENTER
To move between commands within a group. Move forward or backward through the commands in order.	Tab or SHIFT + Tab
If selected command is a list (such as the Font list), use to move between items in the command list	Up Arrow or Down Arrow

Navigate & Open Objects

To do this	Press
Open selected table, query, form, report, macro, or module in Design view	CTRL + ENTER
Display Immediate window in Visual Basic Editor	CTRL + G
Open selected table or query in Datasheet view	ENTER
Open selected form or report	ENTER
Run selected macro	ENTER

Navigate Between a Main Form and a Sub Form

To do this	Press
Exit sub form and move to previous field in main form or previous record	CTRL + SHIFT + Tab
Exit sub form and move to next field in master form or next record	CTRL + Tab

To do this	Press
Enter sub form from following field in main form	SHIFT + Tab
Enter sub form from preceding field in main form	Tab

Navigate Between Forms Greater Than 1 Page

To do this	Press
Move down 1 page; at end of record, moves to equivalent page on next record	Page Down
Move up 1 page; at end of record, moves to equivalent page on previous record	Page Up

Navigate Between Print Preview and Layout Preview

To Do This	Key
Move to the page number box	ALT + F5 (type page number and press ENTER)
Cancel Print Preview or Layout Preview	C or Esc
Move to bottom of page	CTRL + Down Arrow
Move to lower-right corner of page	CTRL + End
Move to upper-left corner of page	CTRL + Home
Open Print dialog box from Print (for datasheets, forms, and reports)	CTRL + P
Move to top of page	CTRL + Up Arrow
Move to right edge of page	End

To Do This	Key
Move to left edge of page	Home
Scroll down 1 full screen	Page Down
Scroll up 1 full screen	Page Up
Open Page Setup dialog box (for forms and reports)	S
Scroll down in small increments	Down Arrow
Scroll to left in small increments	Left Arrow
Scroll to right in small increments	Right Arrow
Scroll up in small increments	Up Arrow
Zoom in or out on part of page	Z

Navigate in Datasheet View

To do this	Press
Move to current field in last record	CTRL + Down Arrow
Move to last field in last record	CTRL + End
Move to first field in first record	CTRL + Home
Move to current field in first record	CTRL + Up Arrow
Move to last field in current record	End
Go to specific record	F5, then type record number and press ENTER

To do this	Press
Move to first field in current record	Home
Move to previous field	SHIFT + Tab, or Left Arrow
Move to current field in next record	Down Arrow
Move to next field	Tab or Right Arrow
Move to current field in previous record	Up Arrow

Navigate in Design View

To Do This	Key
Invoke Field List pane in form or report (if Field List pane is already open, focus moves to Field List pane)	ALT + F8
Move selected control down by a pixel (irrespective of the page's grid) NOTE: For controls in a stacked layout, this switches the position of the selected control with the control directly below it, unless it is already the lowermost control in the layout.	CTRL + Down Arrow
Move selected control to left by a pixel (irrespective of page's grid)	CTRL + Left Arrow
Move selected control to right by a pixel (irrespective of the page's grid)	CTRL + Right Arrow
Move selected control up by a pixel (irrespective of the page's grid)	CTRL + Up Arrow

To Do This	Key
NOTE: For controls in a stacked layout, this switches the position of the selected control with the control directly above it, unless it is already the uppermost control in the layout.	
Cut the selected control and copy it to the Clipboard	CTRL + X
Open or close property sheet	F4 or ALT + ENTER
Switch between upper and lower portions of a window (Design view of queries, macros, and the Advanced Filter/Sort window)	F6
NOTE: Use F6 when the Tab key does not take you to the section of the screen you want.	
Toggle forward between design pane, properties, Navigation Pane, ribbon, and Zoom controls (Design view of tables, forms, and reports)	F6
Increase height of selected control (from the bottom) by a pixel	SHIFT + Down Arrow
Switch from the Visual Basic Editor to form or report Design view when you have code module open	SHIFT + F7
Switch from a control's property sheet in form or report Design view to design surface without changing control focus	SHIFT + F7
Decrease width of selected control (to left) by a pixel	SHIFT + Left Arrow

To Do This	Key
NOTE: For controls in a stacked layout, this decreases the width of the whole layout.	
Increase width of selected control (to right) by a pixel NOTE: For controls in a stacked layout, this increases the width of the whole layout.	SHIFT + Right Arrow
Decrease height of selected control (from bottom) by a pixel	SHIFT + Up Arrow
Move selected control down by a pixel along page's grid NOTE: For controls in a stacked layout, this switches the position of the selected control with the control directly below it, unless it is already the lowermost control in the layout.	Down Arrow
Move selected control to left by a pixel along page's grid	Left Arrow
Move selected control to right by a pixel along page's grid	Right Arrow
Move selected control up by a pixel along page's grid NOTE: For controls in a stacked layout, this switches the position of the selected control with the control directly above it, unless it is already the uppermost control in the layout.	Up Arrow

Navigate in Workspace

To Do This	Key
Switch between Visual Basic Editor and previous active window	ALT + F11
Go to Navigation Pane Search box (if focus is already on Navigation Pane)	CTRL + F
Maximize or restore selected window.	CTRL + F10
Switch to next or previous database window	CTRL + F6 or CTRL + SHIFT + F6
	CTRL + F8
Turn on Resize mode for active window when not maximized	(press the arrow keys to resize the window, and then, to apply the new size, press ENTER)
Close active database window	CTRL + W or CTRL + F4
Restore selected minimized window when all windows are minimized	ENTER
Show or hide Navigation Pane	F11
Switch to next or previous pane in workspace	F6 or SHIFT + F6

Navigate in Form View

To do this	Press
Move to last control on form and set focus in last record	CTRL + End
Move to first control on form and set focus in first record	CTRL + Home
Move to current field in next record	CTRL + Page Down
Move to current field in previous record	CTRL + Page Up
Move to last control on form and remain in current record	End
Go to specific record	F5, then type record number and press ENTER
Move to first control on form and remain in current record	Home
Move to previous field	SHIFT + Tab
Move to next field	Tab

Navigate to Another Screen of Data in Datasheet View

To do this	Press
Move right 1 screen	CTRL + Page Down
Move left 1 screen	Move right one screen
Move up 1 screen	Page Down

To do this	Press
Move down 1 screen	Page Up

Open & Save

To Do This	Key
Open folder 1 level above selected folder	Backspace
Open new database	CTRL + N
Open existing database	CTRL + O or CTRL + F12
Save database object	CTRL + S or SHIFT + F12
Delete selected folder or file	Delete
Open selected folder or file	ENTER
Open Save As dialog box	F12 or ALT + F+S
Open Look in list	F4 or ALT + I
Display shortcut menu for selected item such as a folder or file	SHIFT + F10
Move backward through options	SHIFT + Tab
Move forward through options	Tab

Print Database Options

To do this	Press
Cancel Print Preview or Layout Preview	C or Esc
Print current or selected object	CTRL + P
Return to your database from the Backstage	Esc
Open the Print dialog box from Print Preview	P or CTRL + P
Open Page Setup dialog box from Print Preview	S

Refresh Fields with Current Data

To do this	Press
Recalculate fields in window	F9
Refresh contents of a Lookup field list box or combo box	F9
In a sub form, this re-queries underlying table only for sub form	SHIFT + F9

Select & Move a Column in Datasheet View

To do this	Press
Turn on Move mode	CTRL + SHIFT + F8 (then press Right Arrow key or Left Arrow key to move selected column(s) to right or left)

To do this	Press
Select current column or cancel column selection (Navigation mode only)	CTRL + Spacebar
Extend selection 1 column to left, if current column is selected	SHIFT + Left Arrow
Extend selection 1 column to right, if current column is selected	SHIFT + Right Arrow

Select Field or Record

To do this	Press
Change size of selection by 1 word to left	CTRL + SHIFT + Left Arrow
Change size of selection by 1 word to right	CTRL + SHIFT + Right Arrow
Change size of selection by 1 character to left	SHIFT + Left Arrow
Change size of selection by 1 character to right	SHIFT + Right Arrow
Switch between selecting current record and first field of current record, in Navigation mode	SHIFT + Spacebar
Extend selection to previous record, if current record selected	SHIFT + Up Arrow
Select next field	Tab

Undo Changes

To do this	Press
Undo typing	CTRL + Z or ALT + Backspace
Undo changes in current field or current record (if both the current field and current record have been changed, press Esc twice to undo changes, first in the current field and then in the current record)	Esc

Use a Combo or List Box

To do this	Press
Open combo box	F4 or ALT + Down Arrow
Refresh contents of Lookup field list box or combo box	F9
Move down 1 page	Page Down
Move up 1 page	Page Up
Move down 1 line	Down Arrow
Exit combo box or list box	Tab

Use a Diagram Pane

To do this	Press
Remove selected table, view, or function, or join line from query	Delete
Remove selected data column from query output	Spacebar or Minus Sign (-)
Choose selected data column for output	Spacebar or Plus Sign (+)
Move between columns in a table, view, or function	Up, Down, Left, Right Arrow Keys
Move among tables, views, and functions (and join lines, if available)	Tab or SHIFT + Tab

Use a Grid Pane

To do this	Press
Move to last row in current column	CTRL + Down Arrow
Move to bottom right cell	CTRL + End
Move to upper-left cell in visible portion of grid	CTRL + Home
Select entire grid column	CTRL + Spacebar
Move to first row in current column	CTRL + Up Arrow
Paste text from the Clipboard (in Edit mode)	CTRL + V

To do this	Press
Cut selected text in cell and place it on the Clipboard (in Edit mode)	CTRL + X
Clear selected contents of cell	Delete
Clear all values for a selected grid column	Delete
Toggle between Edit mode and cell selection mode	F2
Toggle the check box in Output column NOTE: If multiple items are selected, pressing this key affects all selected items.	Spacebar
Move among cells	The arrow keys, the Tab key, or SHIFT + Tab
Toggle between insert and overtype mode while editing in a cell	INSERT
Move in a drop-down list	Up Arrow or Down Arrow

Use Dialog Boxes

To do this	Press
Open selected drop-down list box	ALT + Down Arrow
Select option, or select or clear the check box by the letter underlined in the option name	ALT + letter key

To do this	Press
Move between options in selected drop-down list box, or move between options in a group of options	Arrow keys
Switch to the previous tab in a dialog box	CTRL + SHIFT + Tab
Switch to next or previous tab in a dialog box	CTRL + Tab or CTRL + SHIFT + Tab
Perform action assigned to default button in dialog box	ENTER
Cancel command and close dialog box	Esc
Open list if it is closed and move to option in list	First letter of an option in a drop-down list
Perform action assigned to selected button; select or clear check box	Spacebar
Move to next or previous option or option group	Tab or SHIFT + Tab

Use Menus

To do this	Press
Close the visible menu and submenu at the same time	Alt
Show the program icon menu (on the program title bar), also known as the control menu	ALT + Spacebar
Show Key Tips	Alt or F10
Move to top or bottom of selected gallery list	CTRL + Home or CTRL + End

To do this	Press
Close visible menu; or, with submenu visible, close only submenu	Esc
Select first or last command on menu or submenu	Home or End
Scroll up or down in selected gallery list	Page Up or Page Down
Open shortcut menu or open drop-down menu for selected gallery item	SHIFT + F10
Open selected menu, or perform action assigned to selected button	Spacebar or ENTER
With menu or submenu visible, select next or previous command	Down Arrow or the Up Arrow key
Select menu to left or right; or, when a submenu is visible, switch between main menu and submenu	Left Arrow or Right Arrow

Use Property Sheets

To do this	Press
Display a property sheet in Design view	ALT + ENTER
Toggle backward between tabs when a property is selected	CTRL + SHIFT + Tab
Toggle forward between tabs when a property is selected	CTRL + Tab
Show or hide property sheet	F4

To do this	Press
Move among choices in control selection drop-down list 1 page at a time	Page Down or Page Up
With a property selected, move up 1 property on tab; or if already at top, move to tab	SHIFT + Tab
Move among choices in control selection drop-down list 1 item at a time	Down Arrow or the Up Arrow key
Move among property sheet tabs with a tab selected, but no property selected	Left Arrow or Right Arrow
Move to property sheet tabs from control selection drop-down list	Tab
With a property already selected, move down 1 property on a tab	Tab

Use Field List Pane with a Form or Report in Design View or Layout View

To do this	Press
Show or hide Field List pane	ALT + F8
Add selected field to form or report detail section	ENTER
Move between upper and lower panes of Field List	Tab
Move up or down Field List pane	Up Arrow or Down Arrow

Use Wizards

To do this	Press
Move to previous page of a wizard	ALT + B
Complete wizard	ALT + F
Move to next page of wizard	ALT + N
Toggle focus between sections (header, body, footer) of wizard	F6
Toggle focus forward between controls in wizard	Tab

Work and Move in Tables

To do this	Press
Move down 1 paragraph	CTRL + Down Arrow
Move to end of a text box	CTRL + End
Move to beginning of text box	CTRL + Home
Insert tab in cell	CTRL + Tab
Move up 1 paragraph	CTRL + Up Arrow
Move to end of a line	End
Start new paragraph	ENTER
Move to beginning of line	Home
Repeat last Find action	SHIFT + F4
Move to preceding cell	SHIFT + Tab

To do this	Press
Move to next row	Down Arrow
Move down 1 line	Down Arrow
Move 1 character to left	Left Arrow
Move 1 character to right	Right Arrow
Move to next cell	Tab
Add new row at bottom of table	Tab at end of last row
Move to preceding row	Up Arrow
Move up 1 line	Up Arrow

Work in Design, Layout, or Datasheet View

To do this	Press
Switch from Visual Basic Editor back to Form or Report Design view	ALT + F11
Toggle backward between views when in a table, query, form, or report NOTE: If additional views are available, successive keystrokes move the focus to the previous view.	CTRL + Left Arrow or CTRL + . (period) NOTE: CTRL + . (period) does not work under all conditions with all objects.
Toggle forward between views when in a table, query, form, or report	CTRL + Right Arrow or CTRL + , (comma)

To do this	Press
NOTE: If additional views are available, successive keystrokes move the focus to the next available view.	
Exit Navigation mode and return to Edit mode in a form or report	Esc
Switch between Edit mode (with insertion point displayed) and Navigation mode in a datasheet	F2
Switch to property sheet (Design view and Layout view in forms and reports)	F4 or ALT + ENTER
Switch to Form view from Form Design view	F5
Switch between upper and lower portions of a window (Design view of queries, macros, and the Advanced Filter/Sort window)	F6
Cycle through field grid, property sheet, field properties, Navigation Pane, Quick Access Toolbar, and Key Tips on ribbon (Design view of tables)	F6
Open Choose Builder dialog box from a selected control on form or report (Design view only)	F7
Open Visual Basic Editor from selected property in property sheet for form or report	F7

Work with Sub Data Sheets

To do this	Press
Go to a specific record in a sub-datasheet (Move focus from sub-datasheet to record number box)	ALT + F5, (type record number and press ENTER)
Move from the datasheet to the record's sub-datasheet	CTRL + SHIFT + Down Arrow
Exit sub-datasheet and move to last field of previous record in datasheet	CTRL + SHIFT + Tab
Collapse sub-datasheet	CTRL + SHIFT + Up Arrow
Exit sub-datasheet and move to first field of next record in datasheet	CTRL + Tab
Enter sub-datasheet from first field of following record in datasheet	SHIFT + Tab
From datasheet, bypass sub-datasheet and move to next record in datasheet	Down Arrow
Enter sub-datasheet from last field of previous record in datasheet	Tab
From last field in sub-datasheet enter next field in datasheet	Tab
From datasheet, bypass sub-datasheet and move to previous record in datasheet	Up Arrow

To do this	Press
Go to a specific record in a sub-datasheet (Move focus from sub-datasheet to record number box)	ALT + F5, (type record number and press ENTER)

www.ingramcontent.com/pod-product-compliance
Lightning Source LLC
Chambersburg PA
CBHW081458200326
41518CB00015B/2305